David W. Taylor

The Speed and Power of Ships

A Manual of Marine Propulsion

David W. Taylor

The Speed and Power of Ships

A Manual of Marine Propulsion

ISBN/EAN: 9783954272723
Erscheinungsjahr: 2013
Erscheinungsort: Bremen, Deutschland

© maritimepress in Europäischer Hochschulverlag GmbH & Co. KG, Fahrenheitstr. 1, 28359 Bremen. Alle Rechte beim Verlag und bei den jeweiligen Lizenzgebern.

www.maritimepress.de | office@maritimepress.de

Bei diesem Titel handelt es sich um den Nachdruck eines historischen, lange vergriffenen Buches. Da elektronische Druckvorlagen für diese Titel nicht existieren, musste auf alte Vorlagen zurückgegriffen werden. Hieraus zwangsläufig resultierende Qualitätsverluste bitten wir zu entschuldigen.

THE SPEED AND POWER
OF SHIPS

A MANUAL OF MARINE PROPULSION

BY

D. W. TAYLOR, E.D., D. Sc., L.L.D.

REAR ADMIRAL (C.C.), U. S. N., RETIRED

HONORARY VICE-PRESIDENT SOCIETY OF NAVAL ARCHITECTS AND
MARINE ENGINEERS, MEMBER INSTITUTION
OF NAVAL ARCHITECTS

VOL. I. TEXT
VOL. II. TABLES AND PLATES

NEW YORK
JOHN WILEY & SONS, INC.
LONDON: CHAPMAN & HALL, LIMITED

PREFACE

The intention of this work is to treat in a consistent and connected manner, for the use of students, the theory of resistance and propulsion of vessels and to give methods, rules and formulæ which may be applied in practice by those who have to deal with such matters. The contents are based largely upon model experiments, such as were initiated in England nearly half a century ago by Mr. William Froude and are now generally recognized as our most effective means of investigation in the field of resistance and propulsion. At the same time care has been taken to point out the limitations of the model experiment method and the regions where it ceases to be a reliable guide.

During the years that the author has directed the work of the U. S. Experimental Model Basin many results obtained there have been published in the Transactions of the Society of Naval Architects and Marine Engineers and elsewhere, so, naturally, the experiments at the U. S. Model Basin have been made large use of wherever applicable. It will be found, however, that they are in substantial agreement with the many published results of the work of other experimental establishments of this kind.

Although the coefficients and constants for practical application are mainly derived from the author's experience at the Model Basin and elsewhere, and are necessarily general in their nature, endeavor has been made wherever possible to develop formulæ and methods in such a manner that naval architects and engineers using the book may, if they wish, adopt their own constants derived from their special experience.

For instance, by the methods given it will be found possible to estimate closely the effective horse-power of a vessel having the form of what I have called the Standard Series, but it will also be found possible, by the same methods, to determine with fair accuracy the variation of resistance with changes of dimensions, etc., of vessels upon almost any lines for which a naval architect may have reliable data, and which, on account of satisfactory past results, or for other reasons, he may wish to use.

The science of Naval Architecture is not yet developed to a point where our knowledge of resistance and propulsion is complete. While the author naturally hopes that this volume will at least partially bridge some of the gaps hitherto existing, much work remains to be done, and in a number of places attention is called to the need of further investigation of various questions. While we know something, for instance, in a qualitative way of the effect of shallow water upon resistance, information which would enable us to solve satisfactorily many problems arising in this connection is lacking, and apparently can be obtained only by much experimental investigation. When dealing with questions of wake and thrust deduction we are not yet upon firm ground, and it is to be hoped that the excellent work recently done by Luke in this connection will soon be supplemented by even more extensive investigations.

D. W. TAYLOR.

WASHINGTON, D.C., *July,* 1910.

CONTENTS

CHAPTER I
Preliminary and General

SECTION		PAGE
1.	STREAM LINES	1
2.	TROCHOIDAL WATER WAVES	10
3.	THE LAW OF COMPARISON	26
4.	WETTED SURFACE	36
5.	FOCAL DIAGRAMS	48
6.	THE DISTURBANCE OF THE WATER BY A SHIP	50

CHAPTER II
Resistance

7.	KINDS OF RESISTANCE	57
8.	SKIN RESISTANCE	58
9.	EDDY RESISTANCE	66
10.	WAVE RESISTANCE	73
11.	AIR RESISTANCE	82
12.	MODEL EXPERIMENT METHODS	87
13.	FACTORS AFFECTING RESISTANCE	90
14.	PRACTICAL COEFFICIENTS AND CONSTANTS FOR SHIP RESISTANCE	98
15.	SQUAT AND CHANGE OF TRIM	108
16.	SHALLOW WATER EFFECTS	112
17.	ROUGH WATER EFFECTS	121
18.	APPENDAGE RESISTANCE	123

CHAPTER III
Propulsion

19.	NOMENCLATURE, GEOMETRY AND DELINEATION OF PROPELLERS	128
20.	THEORIES OF PROPELLER ACTION	136
21.	LAW OF COMPARISON APPLIED TO PROPELLERS	150
22.	IDEAL PROPELLER EFFICIENCY	153
23.	MODEL EXPERIMENTS — METHODS AND PLOTTING RESULTS	155
24.	MODEL PROPELLER EXPERIMENTS — ANALYSIS OF RESULTS	158
25.	PROPELLER FEATURES INFLUENCING ACTION AND EFFICIENCY	160
26.	PRACTICAL COEFFICIENTS AND CONSTANTS FOR FULL-SIZED PROPELLERS DERIVED FROM MODEL EXPERIMENTS	175
27.	CAVITATION	182

SECTION		PAGE
28.	Wake Factor, Thrust Deduction and Propeller Suction	195
29.	Obliquity of Shafts and of Water Flow	211
30.	Strength of Propeller Blades	216
31.	Design of Propellers	241
32.	Paddle Propulsion	254
33.	Jet Propulsion	260

CHAPTER IV

Trials and Their Analysis

34.	Measured Courses	262
35.	Conduct of Speed and Power Trials	264
36.	Analysis of Trial Results	279

CHAPTER V

The Powering of Ships

37.	Powering Methods Based upon Surface	291
38.	The Extended Law of Comparison	295
39.	Standard Series Method	300

THE
SPEED AND POWER OF SHIPS

CHAPTER I

Preliminary and General

1. Stream Lines

1. Assumptions Made. — The consideration of stream lines or lines of flow will be restricted mainly to the case of the motion of liquid past a solid. It is sufficient for present purposes to define a liquid as a fluid which is incompressible, or virtually so, such as water.

The difficulties in the way of adequate mathematical determination of the motion of liquids past solids such as ships have hitherto been found insuperable. The mathematics of the motion of liquids is complicated; even the simple cases which can be dealt with mathematically require assumptions which are far from actual conditions in practice. Thus, when considering the motion of solids through a liquid, or what is the same thing mathematically, the motion of a liquid past solids, it is assumed that the liquid is "perfect" or has no viscosity and that the solid is frictionless, that is to say, that the liquid can act upon the solid only by pressure which must at each point be normal to the surface. In most cases that are dealt with mathematically, it is further assumed that the fluid or liquid extends to an infinite distance from the solid.

2. Steady Motion Formula. — We cannot deal satisfactorily with problems of resistance by mathematical analysis, but in spite of the somewhat artificial assumptions involved, the results of mathematical analysis applied to a perfect liquid are of interest and value as they indicate tendencies and have large qualitative bearing upon the phenomena of the motion of water past ships.

One mathematical conclusion in this connection is particularly valuable. It is known as the steady motion formula and is as follows:

$$\frac{p}{w} + \frac{v^2}{2g} + z = h.$$

In the above formula, p denotes pressure of the liquid per unit area, w denotes weight per unit volume, v denotes velocity of flow in units of length per second, g acceleration due to gravity in units of length per second, z denotes height above a fixed level and h is a constant for each stream line, being called the head. It is usually convenient to express p in pounds per square foot, w in pounds per cubic foot, v and g in feet per second, z and h in feet.

The above formula applies to the steady motion of an infinite mass of perfect liquid. For such liquid the value of h is constant for all particles passing a point fixed in the liquid. These particles form a continuous line called a stream line, and in steady motion, no matter how many twists and turns the stream line takes, the above formula applies to its pressure, velocity and elevation at every point. It will be observed that contrary to what might at first be thought, the greater the velocity at a point of the stream line the less the pressure, and vice versa. That is to say, if a stream of perfect liquid flows in a frictionless pipe of gently varying section, the pressure increases as the size of the pipe increases and decreases as the size of the pipe decreases. This is demonstrable in the case of flow through pipes, although it is necessary to have the changes of section very gradual in order to obtain the smooth continuous motion to which alone the steady motion formula is applicable.

3. **Application of Steady Motion Formula to Ships.** — The steady motion formula applies to the motion of a liquid, including motion past a solid at rest. In the case of ships, we are interested in the motion of a solid through a liquid at rest. The two cases are, however, as already stated, mathematically interchangeable. Suppose we have a ship moving uniformly through still water which extends indefinitely ahead and astern. If we suppose both ship and water given the same velocity, equal and opposite to the velocity of the ship in the still water we have the ship at rest and the water flowing past it. The mutual reactions between ship

and water are identical whether we have the ship moving through still water or the water flowing past the fixed ship. To the latter case, however, the steady motion formula applies if we neglect friction and the mathematical treatment is much easier.

If the ship is in a restricted channel so shallow and narrow that the area of the midship section of the ship is an appreciable fraction of the area of the channel section, the steady motion formula teaches us that with the water flowing past the fixed ship there will be abreast the central portion of the ship where the channel area is diminished an appreciable increase in velocity of flow and reduction of pressure.

The surface being free, reduction of pressure would result in depression of surface. Passing to the case of the ship moving through the channel we would infer that the water is flowing aft abreast the central portion of the ship and that there is a depression in this vicinity.

This, as a matter of fact, occurs in all cases, but in open water the motions are not so pronounced, and it is seldom possible to detect them by the eye. In a constricted channel, however, it is generally easy to detect the depression abreast the ship since it extends to the banks. If these are sloping the depression shows more plainly than it does against vertical or steep banks.

There might be quoted many other illustrations of the validity of the steady motion formula taken from phenomena of experience. There is no doubt of its general validity within certain limits as regards motion of water around solids, but in considering any particular case it should not be applied regardless of its limitations.

4. Failure of Steady Motion Formula. — The steady motion formula assumes frictionless motion. Water is not frictionless, but its friction is not sufficiently great in the majority of cases to seriously affect steady motion directly.

The main failure of the steady motion formula as regards practical cases is in connection with the transformation of pressure into velocity and vice versa. Neglecting variations of level the steady motion formula is $\frac{p}{w} + \frac{v^2}{2g} =$ a constant. By the formula the greater the velocity the less the pressure, and if the velocity be made sufficiently great the pressure must become negative. Now, negative pressure would be a tension, and liquids are physically incapable of standing a tension. Hence, when the case is such that the steady motion formula would give a tension the motion that would be given by the steady motion formula becomes impossible and the formula fails. In practice, in such a case, instead of steady motion we have eddying, disturbed motion. In fact, in actual liquids, when the motion is such as to cause a reduction of pressure, eddying generally makes its appearance some time before the pressure becomes zero. But for moderate variations of pressure we find for actual liquids pressure transformed into velocity according to the steady motion formula with great accuracy. The transformation of velocity into pressure, however, according to the steady motion formula, without loss of energy, is not common in practice. For instance, experiments at the United States Model Basin have shown that air will pass through converging conical pipes with practically no loss of head except that due to friction of the pipe surface. But when passing through diverging cones, even when the taper is but one-half inch of diameter per foot of length, there is material loss of head beyond that due to friction. It appears reasonable to suppose that the difficulties found in converting velocity of actual fluids into pressure without loss of energy are connected with the friction of the actual fluids, both their internal friction or viscosity and their friction against the pipes or vessels containing them.

To sum up, we appear warranted in concluding that in flowing water pressure will be transformed into velocity according to the steady motion formula with little or no loss of energy in most cases, provided the pressure is not reduced to the neighborhood of zero, and that velocity will be transformed into pressure but with a loss of energy dependent upon the conditions.

It is evident that if the total head or average pressure is great, given variations of pressure and velocity can take place with closer approximation to the steady motion formula than if the total head be small.

5. **Sink and Source Motion.** — The mathematics of fluid motion or hydrodynamics being somewhat complicated will not be gone

into here, but results will be given in a few of the simplest cases which are of interest and have practical bearing. Suppose we have liquid filling the space between two frictionless planes which are very close together. The motion will be everywhere parallel to the planes, and hence will be uniplanar or in two dimensions only. Suppose now that liquid is being continually introduced between the planes at some point. It will spread radially at an equal rate in every direction. The point of introduction of the liquid is called a "source." Fig. 1 indicates the motion, S being the source. If liquid were being abstracted at S the motion at every point would be directly opposite that shown in Fig. 1 and S would be what is called a "sink." The sink and source motion is not physically possible because the steady motion formula applies, and for velocity and pressure finite at a distance from S the velocity at S would be infinite. But it will be seen presently that the mathematical concept of sinks and sources has a bearing upon possible motions. Suppose that instead of a single source or sink we have in Fig. 2 a source at A and a sink of equal strength at B. Liquid is being withdrawn at B at the same rate at which it is being introduced at A and in time every particle introduced at A must find its way out at B. The motion being steady the paths followed are stream lines. These paths are arcs of circles. A number of these circular arcs are indicated in Fig. 2. They are so chosen that the " flow " or quantity of fluid passing between each pair of circles is the same. Adjacent to the line connecting the sink and source the path is direct, the velocity great and the circles close together. As we leave this line the path followed from source to sink is circuitous, the velocity low and the spacing of the circles greater and greater.

6. Sink and Source Motion Combined with Uniform Stream. — Suppose, now, that the liquid in which the source is found is not at rest but is flowing with constant speed from right to left. Fig. 3 shows the result of the injection of a source into such a uniform stream. In this case we have a curve of demarcation DDD separating the liquid which comes rom the source and the other liquid. No liquid crosses this curve. Now, the motion being frictionless it makes no difference whether DDD is an imaginary line in the moving liquid or the boundary of a frictionless solid. Hence if in a uniform stream we put a frictionless solid of the shape DDD the motion outside of it will be the same as in Fig. 3. This motion will be completely possible if we could have a frictionless solid like DDD, since we no longer have the source with its impossible conditions as regards velocity and pressure.

In Fig. 3 DDD extends to infinity. Suppose, now, in a uniform stream we put a sink and a source of equal strength as at A and B in Fig. 4. The direction of flow of the uniform stream is supposed parallel to AB. In this case the closed oval curve CCC separates the liquid which appears at the source and disappears at the sink from the liquid of the uniform stream. Hence, if a frictionless solid of the shape of CCC took the place of the liquid inside the oval the motion of the stream outside would be unchanged.

Of course the shape and dimensions of CCC would vary with the relative strengths of source and sink and velocity of stream. Instead of one source and one sink we may distribute a number along the line AB enabling us to modify the shape and proportions of the line of demarcation CCC. The author (see Transactions of the Institution of Naval Architects for 1894 and 1895) has extended this method to cover the case of an infinite number of infinitely small sources and sinks, thus enabling us to determine lines of demarcation or stream forms both in plane and solid motions, closely resembling actual ships' lines. Not only the stream forms but also the velocities and pressures along them can be determined, but the process is laborious and has not so far been given sufficient practical application to warrant following further here.

The closed ovals due to a source and a sink in a uniform stream somewhat resemble ellipses as appears from Fig. 4.

7. Flow in Two Dimensions in Practice. — While to reduce the motion to one plane or two dimensions, the assumption was made that it took place between two frictionless parallel planes so close together that the space between them practically constituted a single plane, it should be pointed out that motion practically identical with plane motion occurs in practice. Suppose we have a body of cylindrical type of infinite length moving in some direction perpendicular to its axis. The motion past will be identical in all planes perpendicular to the axis.

The motion past an actual body of cylindrical type whose length though not infinite is great compared with its transverse dimensions will, over a great portion of the length, be practically the same as if the length were infinite. A propeller strut is a case in point. Ideal plane flow has direct practical bearing upon the motion past such fittings.

8. Stream Lines past Elliptic Cylinders. — One general case of uniplanar motion that has been solved mathematically is that of an elliptic cylinder moving parallel to either axis in an infinite mass of liquid. The circle is a special case and a plane lamina is another special case where one axis of the ellipse is zero. The general mathematical formulæ expressing the motion of an elliptic cylinder through liquid may be referred to in Lamb's "Hydrodynamics," edition of 1906, Article 71. They do not give directly the stream lines past an elliptic cylinder but the latter can be deduced from them. Figs. 5 to 15 show plane stream lines or lines of flow past various types of elliptic cylinders. The lines in the first quadrant only are shown as they are symmetrical in the other three. The proportions of the ellipses are given, the semi-major axis being always taken as unity. Fig. 10 shows flow around a circular cylinder and Fig. 15 flow past a plane lamina of indefinite length and unit half breadth. The flow around a lamina is, however, impossible since the formula would require an infinite velocity around the edges, or, as indicated in Fig. 15, the stream line spacing in the immediate vicinity of the edge would become infinitely narrow.

9. Pressure Variations around Elliptic Cylinders. — Figs. 16 and 17 give some idea of variation of pressure along the central stream line and around the surface of the cylinders. A particle approaching a cylinder along the axis steadily loses velocity and gains pressure until it comes to rest against the cylinder when its pressure is increased by the total velocity head of the undisturbed stream. The particle then starts around the cylinder, rapidly gaining velocity and losing pressure until at a point where it has moved but a short distance around the cylinder it has regained the velocity and returned to the pressure it had in the undisturbed stream. The velocity then continues to increase and the pressure falls as shown until the particle is abreast the center of the cylinder when the velocity is at a maximum and the pressure at a minimum.

Figs. 16 and 17 show negative pressures but these are only relatively negative. For convenience the diagrams are drawn as if the pressure in the undisturbed stream were zero. The actual pressure in any case is the pressure of the figure with the pressure in the undisturbed stream added. Bearing in mind also that in each figure the unit of pressure is the pressure head due to the velocity of the undisturbed stream, or the velocity head of the stream, Figs. 16 and 17 shed a good deal of light upon the effect of variation of proportions. Thus, for an ellipse one-tenth as wide as long, the maximum reduction of pressure abreast the center is about one-fifth the velocity head. For the ellipse four-tenths as wide as long, the maximum reduction is nearly the velocity head. For the ellipse as wide as long (the circle), the reduction is three times the velocity head. For the ellipse two and one-half times as wide as long, the reduction is over eleven times the velocity head, and for the ellipse five times as wide as long, the reduction is thirty-five times the velocity head and about one hundred and seventy-five times the reduction for the ellipse one-tenth as wide as long.

The velocity head being proportional to the square of the speed, the reduction in or increase of pressure at every point is proportional to the square of the speed, and hence if any of the cylinders were pushed to a high enough speed the reduction of pressure abreast the center would equal the original pressure in the undisturbed stream, and hence the pressure abreast its center would reduce to zero resulting in eddying. But eddying would appear in the case of an actual cylinder long before the pressure abreast the center became zero. For the excess velocity amidships would not be fully converted into excess pressure on the rear of the cylinder as required for perfect stream motion, and eddying would show itself aft.

10. **Disturbance Abreast Cylinder Centers.** — It is evident from Figs. 16 and 17 that in the case of a cylinder moving through still water the maximum sternward velocity of the water at any point of the cylindrical surface is abreast the center of the cylinder. It is also true that for motion parallel to the axis of x the greatest

sternward velocity for any value of y is on the axis of y. It is of interest to trace the variation of velocity as we pass along the axis of y. Fig. 18 shows sections of seven types of cylinders ranging from the flat plate, No. 1, which is all breadth, to the circle, No. 4, and the ellipse five times as long as wide, No. 7. They all have unit half breadth on the axis of y and are supposed to move with velocity V parallel to the axis of x.

Fig. 18 shows also curves of sternward velocity u of the water as we pass out from the cylinder along the axis of y expressed as a fraction of the speed of advance of the cylinder. It is seen that the long cylinder causes the minimum disturbance at the surface of the cylinder where $y = 1$, but the maximum beyond $y = 4$. Fig. 18 shows markedly the very great variations of disturbance in the vicinity of the cylinder with variation of ratio of breadth to length. The areas of all the curves of Fig. 18 are the same, being equal to $V \times$ (half breadth). The dotted square in the figure shows this area.

11. **Tracks of Particles.** — While Figs. 5 to 15 show stream lines or flow past the cylinders, they give little idea of the paths followed by particles of water when a cylinder is moved through water initially at rest.

Rankine gave, many years ago, the differential equation to these paths for the motion of a circular cylinder, and while this equation cannot be integrated it is possible by graphic methods to determine the resulting paths with ample accuracy.

Fig. 19 shows the paths followed by a few particles at various distances from the axis as a cylinder of the size indicated by the dotted semicircles in the figure passes along the axis from an infinite distance to the right to an infinite distance to the left.

A on each path shows the original position of the particle when the cylinder is at an infinite distance to the right. B, C, D, E and F on the paths of the particles show positions when the cylinder is at $B, C, D, E,$ and F, on the axis as indicated.

The paths are symmetrical, and G denotes the position of each particle when the cylinder has passed to an infinite distance to the left.

Fig. 19 shows the curious result that each particle is shifted ultimately a certain distance parallel to the direction of motion of the cylinder. This could not occur if the cylinder started from rest at a finite distance from the particle, and came to rest within a finite distance of the particle. For such motion the particles must on the average be slightly displaced in a direction opposite to the direction of motion of the cylinder.

12. **Stream Lines around Sphere.** — While there are very few mathematical determinations of stream lines in three dimensions those for the sphere are known and it is of interest to compare them with those for a circular cylinder shown in Fig. 10. The stream lines past a sphere are identical in all planes through the axis parallel to the direction of undisturbed flow.

They are shown in Fig. 20, and in Fig. 21 are shown curves of pressure variation along the horizontal axis and around the sphere, also along the horizontal axis and around a circular cylinder. The curves show as might be expected that the sphere creates less disturbance. This is evidently because the water is free to move in three dimensions around the sphere, while it is restricted to plane motion around the cylinder.

The increase of pressure in front of the sphere is less. There is a sudden rise close to the intersection of axis and sphere. At this point the increase of pressure is the same as in the case of the cylinder, being the pressure head due to the undisturbed velocity. Abreast the center the loss of pressure is one and one-half times that due to the velocity as contrasted with three times the velocity head in the case of the cylinder. In other words, if a sphere is advancing with perfect stream line action through water otherwise undisturbed the water abreast the center is flowing aft with one-half the velocity of advance of the sphere. In the case of the circular cylinder the water abreast its center flows aft with velocity equal to the velocity of advance.

2. Trochoidal Water Waves

1. **Mathematical Waves.** — Ocean waves during a storm are generally confused rather than regular. They are not of uniform height or length from crest to crest, and the crests and hollows extend but comparatively short distances. After a storm, however, the confused motion settles down into rather uniform and

regular swells and the motion approaches that of mathematical waves. For mathematical treatment it is necessary to assume regularity of motion. We may define a series of mathematical waves as an infinite series of parallel infinitely long identically similar undulations advancing at uniform speed in a direction perpendicular to that of their crests and hollows. The constant distance between successive crests is called the length of the waves or the wave length, the distance between the level of the crest and the level of the hollow is called the height of the wave, and the time interval between the passage of successive crests by a fixed point is called the period of the wave.

Mathematical waves are cases of motion in two dimensions, since the motion is identical in all planes perpendicular to the wave crests.

2. Trochoidal Wave Theory. — The most commonly accepted theory of regular wave motion is that called the "trochoidal theory." Its mathematics is too long and difficult to be gone into here, and I shall undertake only to give some of the formulæ and conclusions that have been evolved by the eminent mathematicians who have worked in this field.

Of the British mathematicians who have contributed to the trochoidal theory, Airy and Rankine were especially prominent shortly after the middle of the last century.

By the trochoidal theory, in water of unlimited depth each particle describes at a uniform rate a circular orbit, making one complete revolution per wave period, the radii of the orbits being a maximum for surface particles and decreasing indefinitely with depth.

Referring to Fig. 22 let the wave length be denoted by L and let R be the radius of a circle whose circumference is L. Then $R = \dfrac{L}{2\pi}$. Suppose we locate this circle with its center midway between the levels of crest and hollow and take a point P on the radius at a distance r or $\dfrac{H}{2}$ from the center, H being the wave height. Then, if the circle rolls on the line AB the point P will describe a trochoid giving the outline of the wave surface. This trochoid shows the contour assumed by particles originally at the surface level. Similarly, particles originally at any level below the surface are found along a trochoidal surface having the same diameter of rolling circle but less orbit radius, the radius diminishing indefinitely with depth.

Fig. 23 shows the trochoids at various levels, orbit diameters and contours of lines of particles which in undisturbed water were equally spaced verticals. The cycloid — the limiting trochoid — is shown, but it is not possible for sharp crested waves to appear in practice. They break long before they approach closely the limiting cycloid.

Fig. 23 is for water of unlimited depth. In water of finite depth, by the trochoidal theory each particle describes an elliptical orbit instead of the circular orbit of deep water. Referring to Fig. 24 let $ABCD$ be the "rolling circle" whose perimeter, as before, is equal to the wave length from crest to crest. Let the ellipse $EFGH$ of center the same as the center of the rolling circle be the orbit of the surface particles. Let OP' be the radius of a concentric circle of diameter the same as the major (horizontal) axis of the ellipse. Then, as the rolling circle moves, let the radius OP' revolve with it and the ellipse move horizontally with it without revolving. Draw vertical lines as $P'N$ from the successive positions of P' to meet the ellipse in points such as P. The modified trochoid obtained by joining all points such as P is the surface profile of the wave.

The horizontal and vertical axes of the elliptical orbits are not independent but vary with the depth of water, the depth below the surface, etc.

Thus let a and b denote the horizontal and vertical semi-axes, respectively, of an elliptical orbit whose center is a distance h below the orbit centers of the surface particles. Let $a_0 b_0$ denote the semi-axes of the surface orbit. Let d denote the depth from center of surface orbits to the bottom. Let R denote the radius of the rolling circle and ω the angular velocity with which it must roll to have its center travel at the speed of the wave.

Let L denote the wave length in feet, v the wave speed in feet per second, g the acceleration of gravity and e the base of hyperbolic logarithms. Then the formulæ connecting the above quantities are as follows:

$$a_0 = b_0 \frac{e^{\frac{4\pi d}{L}} + 1}{e^{\frac{4\pi d}{L}} - 1},$$

$$b = b_0 \frac{e^{\frac{2\pi(d-h)}{L}} - e^{\frac{-2\pi(d-h)}{L}}}{e^{\frac{2\pi d}{L}} - e^{\frac{-2\pi d}{L}}},$$

$$a = b_0 \frac{e^{\frac{2\pi(d-h)}{L}} + e^{\frac{-2\pi(d-h)}{L}}}{e^{\frac{2\pi d}{L}} - e^{\frac{-2\pi d}{L}}},$$

$$v = \omega R \quad \text{and} \quad \omega = \sqrt{\frac{b_0}{a_0} \frac{g}{R}}.$$

Whence

$$v^2 = \frac{b_0}{a_0} gR = \frac{b_0}{a_0} \frac{gL}{2\pi},$$

and if T denote the period in seconds

$$T = \frac{L}{v} = \sqrt{\frac{a_0}{b_0} \frac{2\pi L}{g}} = \sqrt{\frac{a_0}{b_0} \frac{4\pi^2 R}{g}}.$$

To pass to the case of indefinitely deep water, we put $d = \infty$. Then $a_0 = b_0 = r_0$, say, and if r denote the radius of the circular orbit at a distance h below the surface orbits, we have

$$a = b = r = r_0 e^{\frac{-2\pi h}{L}}.$$

As before, $v = \omega R$, but

$$\omega = \sqrt{\frac{g}{R}}, \quad v^2 = gR = \frac{gL}{2\pi} \quad \text{and} \quad T = \frac{L}{v} = \sqrt{\frac{2\pi L}{g}}.$$

Substituting for g the value 32.16 and for π its value, we have the following formulæ for deep-water trochoidal waves:

Velocity in feet per second $= v = 2.26 \sqrt{L}$,
Velocity in knots $= V = 1.34 \sqrt{L}$,
Period in seconds $= T = 0.442 \sqrt{L}$.
Length in feet $= .557 V^2 = 5.118 T^2$.

The above rather complicated-looking formulæ express completely the motion under the trochoidal theory.

3. Mechanical Possibility of Trochoidal Waves. — For the motion to be possible it must satisfy,
1. The condition of continuity.
2. The condition of dynamical equilibrium.
3. The boundary conditions.
4. The conditions of formation.

The mathematical investigation of the above conditions is too long and complicated to be given here. The results only can be given. . As regards continuity, it is found that the motion is possible in water of infinite depth, but that in water of finite depth the equation of continuity is not quite satisfied.

As regards dynamical equilibrium, again we find that the motion is not quite possible in finite depth, the pressure at the surface being not quite constant, which it must be from boundary conditions. In infinite depth, however, the pressure as deduced from the trochoidal formulæ is constant along the wave profile and hence the motion is possible.

The only other boundary conditions to be satisfied are those at the bottom, and these are satisfied by the trochoidal formulæ, since they give at the bottom horizontal motion only ($b = 0$) when the water is of finite depth and no motion at all ($r = 0$) when the water is of infinite depth.

Finally, as regards the condition of formation, it is a theorem of hydrodynamics that a perfect liquid, originally at rest, that has been acted upon by natural forces only, cannot show molecular rotation. The trochoidal wave motion involves a slight molecular rotation, and hence falls slightly short of being a possible motion in both finite and infinite depths.

We conclude, then, that trochoidal wave motion falls slightly short of being mathematically possible; but it would require a very small change in the motion to render it possible. This and other considerations which will be pointed out later warrant the adoption of the trochoidal theory as a working approximation.

4. Trochoidal Wave Profiles. — The formulæ already given may be supplemented by those representing the trochoidal contours at various depths. They are $x = R\theta - a \sin \theta$, $y = h - b \cos \theta$, where x is measured horizontally, y is measured vertically down from the surface orbit centers R, a and b have the values already given and θ is angle rolled through by the rolling circle, being $=0$ for an initial condition where the radius of the rolling circle is vertical and its center under the crest of the trochoid. Of course, in deep water $a = b = r$.

Fig. 25 shows the wave surface profiles for three waves, each 300 feet long and 20 feet high, but in three depths of water, namely ∞, 25 feet and 15 feet. These three profiles have the same line of undisturbed water level. It is seen that in each case the orbit center, or mid height of wave, is above the level of the undisturbed water. For deep-water waves the amount of this elevation is $\frac{r_0^2}{2R}$, r_0 being the surface orbit radius. For shallow-water waves it is $\frac{a_0 b_0}{2R}$. The pressure on any trochoidal subsurface for deep-water waves is uniform and the same as the pressure in undisturbed water on the corresponding layer.

For subsurface trochoids the elevation of orbit centers is given by $\frac{r_0^2}{2R} e^{-\frac{2h}{R}}$, where h is the distance of the orbit centers from the level of surface orbit centers.

5. Energy of Trochoidal Waves. — Consider now the energy of waves in deep water. This is partly potential, due to the fact that in wave motion the particles are elevated on the average above their still-water positions, and partly kinetic, due to the velocity with which the particles of water are revolving in their circular orbits.

Let w denote the weight of one cubic foot of water. Then the potential energy of a mass of water one foot wide and one wave length long, i.e., extending from one crest to the next, is

$$E_p = \frac{w\pi}{2} r_0^2 R \left(1 - \frac{r_0^2}{2R^2}\right),$$

where r_0 is surface orbit radius or one-half the wave height.

Now $R = \frac{L}{2\pi}$. Substituting this value we may write

$$E_p = \frac{wr_0^2 L}{4}\left(1 - \frac{2\pi^2 r_0^2}{L^2}\right).$$

In practice, for actual waves $\frac{2\pi^2 r_0^2}{L^2}$ is a small fraction and for most purposes can be ignored. The kinetic energy of the mass of water as above is exactly the same as the potential energy, or if we denote it by E_k,

$$E_k = \frac{w r_0^2 L}{4}\left(1 - \frac{2\pi^2 r_0^2}{L^2}\right).$$

While the potential and kinetic energies of a mass of water in wave motion remain constant, there is constant transmission of energy going on.

Fig. 26 shows a number of positions of a distorted vertical or line of particles originally vertical in still water. During part of the motion, energy is being transmitted across this vertical in the direction in which the wave is traveling and during the rest of the motion it is being transmitted backward. One wave length away is a similar distorted vertical moving in the same way, so there is at no time net gain or loss of energy to a mass of water one wave length long. But the energy transmitted forward across a surface originally a vertical plane is during one wave passage greater than the energy transmitted backward by the quantity $\frac{w r_0^2 L}{4}\left(1 - \frac{2\pi^2 r_0^2}{L^2}\right)$.

This is identical with the kinetic or potential energy of the wave, so that a mass of water extending over one wave length receives from the water behind it and communicates to the water in front of it during the passage of one wave a net amount of energy equal to its kinetic or potential energy.

While this is the net energy transmitted the rate of transmission is much higher during a portion of the wave passage than the average. Thus, if θ is the angle in its orbit from the vertical of the radius r_0 of a surface particle, the rate of transmission of energy through the distorted vertical terminating in the surface particle (see Fig. 26) is given by

$$\frac{dE}{d\theta} = wR\left[R\cos\theta\left(r_0 - \frac{r_0^3}{3R^2}\right) + \frac{r_0^2}{4}\left(1 - \frac{r_0^2}{2R^2}\right)\right].$$

By integrating this between the limits $\theta = 0$ and 2π, we get the expression given above for the net energy transmitted. Fig. 27

shows a curve of rate of transmission of energy for a deep-water wave 300 feet long and 20 feet high. Between 0° and 90° and 270° and 360° there is positive transmission. Between 90° and 270° there is negative transmission. The average rate of transmission is indicated on the figure.

6. Superposition of Trochoidal Waves. — If we superpose two trochoidal wave series of the same length L, and hence the same speed of advance, which are traveling in the same direction with parallel crests a distance a apart, the result is a single series of length L.

If we denote by H_1, H_2 the wave heights of the two components and by H the height of the resultant series, we have

$$H^2 = H_1^2 + H_2^2 + 2 H_1 H_2 \cos \frac{a}{R} = H_1^2 + H_2^2 + 2 H_1 H_2 \cos \frac{2\pi a}{L}.$$

Evidently if $a = 0$, or the crests of the component series are immediately over one another, $\cos \frac{a}{R} = 1$ and $H^2 = (H_1 + H_2)^2$. In this case the wave height of the resultant series is the sum of the component heights. If $a = \pi R$ we have $\cos \frac{a}{R} = -1$ and $H^2 = (H_1 - H_2)^2$. In this case the crest of one component is immediately over the hollow of the other, and the height of the resultant series is the difference of the heights of the components. If in this case $H_1 = H_2$, the components extinguish each other and the resultant is still water.

7. Wave Groups. — A very important deduction from the trochoidal theory is the theory of wave groups. If we superpose two trochoidal systems of equal heights, but slightly different lengths, we have at one point of the resultant series waves of double the height of either component and at another point waves of zero height, since at one point of the series we would have crest superposed on crest and at another point crest superposed on hollow. The resultant series in this case would consist of a number of groups of waves, each with a wave of maximum height in the middle and of heights steadily decreasing ahead and astern of the middle until waves of infinitesimal height or bands of practically still water separate the groups. It can be easily proved from the trochoidal theory that each group will travel as a whole at just half the speed appropriate to the wave length of the original components. The individual waves, however, travel at their natural speed, which is double the group speed. A wave will advance from the rear of a group where its height is infinitesimal and pass through the group, growing until it reaches a maximum at the center of the group and then dwindling as it goes forward until its height again becomes infinitesimal at the front of the group. One can readily start a group of circular waves by dropping a pebble from a bridge into a placid stream. This shows general features somewhat similar to the theoretical trochoidal group. If the reflection in the water of the side of the bridge is distinct a wave can be watched as, first becoming noticeable at the rear, it passes through the group, reaching a maximum height and dying down again as it gets further and further ahead of the center of the group. It will be found, however, that unlike the theoretical trochoidal group, which has similar groups some distance ahead and astern of it, the circular group gets wider and wider from front to rear. If, for instance, at a given time it shows five appreciable waves, it will be seen a little later to show six, then seven, and so on.

8. Applicability of Trochoidal Theory. — Having considered the nature of the motions and the conclusions that can be drawn from the trochoidal wave theory, it is time to consider its applicability to actual water waves. We know that actual waves cannot be exactly trochoidal, and we are not warranted in assuming without some confirmatory evidence that the trochoidal theory gives us waves substantially the same as actual waves. Now, as already pointed out, actual waves are almost never regular, so that a rather rough approximation, mathematically, to the ideal regular waves would, as a rule, resemble them more closely than do the actual waves. Hence, if we find that the trochoidal theory adequately represents the most important feature or features of wave motion we need not be concerned as to minor features.

Stokes has developed a mechanically possible theory of wave motion where the wave profiles are sines and the speed of the wave is not independent of the height, but increases slightly with it.

For waves of ordinary proportions, however, the speed is practically the same as by the less complex trochoidal theory.

It appears, then, that for the proportions occurring in practice trochoidal waves are in substantial agreement with mathematical waves free from their minor mechanical imperfections.

Now, what is the basic feature of trochoidal waves? It seems that it may fairly be said to be the fact that the velocity of advance depends only upon the length from crest to crest and the depth of the water. We have seen that the formula for this velocity is

$$v^2 = \frac{b_0}{a_0}\frac{gL}{2\pi} = \frac{e^{\frac{2\pi d}{L}} - e^{\frac{-2\pi d}{L}}}{e^{\frac{2\pi d}{L}} + e^{\frac{-2\pi d}{L}}} \frac{gL}{2\pi}.$$

Small-scale experiments in tanks, such as those of the Weber Brothers, who published their results in 1825, have given results consistent with the trochoidal theory; but it is obviously desirable to compare the theory with actual full-sized waves, which it is very difficult to do with accuracy.

9. Gaillard's Experimental Investigations of Trochoidal Theory. — Major D. D. Gaillard, U. S. A., in a monograph on Wave Action in Relation to Engineering Structures (Professional Papers, No. 31, Corps of Engineers, U. S. Army), has compared reported speeds of advance and speeds computed by the trochoidal theory in eighty-five cases of ocean waves observed by various people at various places. Of these eighty-five reported velocities, twenty-three were higher than the computed velocities corresponding to the observed length and sixty-two were lower, the average of the whole number being nearly 9 per cent below the average computed velocity. While giving due consideration to the difficulties in the way of accurate observation, the agreement between these observations and the trochoidal theory is certainly not wholly satisfactory.

Fortunately, Major Gaillard gives a further comparison of the trochoidal theory with a large number of observations, taken by himself or under his direction, under conditions favorable to accuracy. These observations were made in 1901 and 1902 in the Duluth, Minn., ship canal and in Lake Superior near the canal.

The canal in question is about 300 feet wide, 26 feet deep,

where the observations were taken, and about 1000 feet long. It connects the harbor of Duluth with Lake Superior, and natural conditions are such that during and after storms, waves often pass squarely into its mouth and on through it. By means of instantaneous photography accurate profiles of waves against the walls, either in the canal or outside, in gently shoaling water, could be determined. The velocity of the waves could also be determined quite accurately, velocity observations being usually taken between stations 300 feet apart. The observations during two years numbered 631 in all. The wave heights varied from 2 to 23 feet, the wave lengths from 45 to 425 feet, and the wave velocities from 9.1 to 33.3 feet per second. The depth of the water varied from 3.3 to 27 feet, though 533 of the observations were taken in the canal 26 feet deep. For these 533 observations the mean observed velocity and the mean velocity as computed from the shallow-water trochoidal formula agreed within less than one-half of one per cent. This is practically exact agreement. For the ninety-eight observations made outside the canal in varying depths the computed velocities averaged nearly 5 per cent more than the observed velocities. Major Gaillard states that conditions and facilities were such that the last series of observations could not be taken with the same degree of accuracy as those on waves inside the canal. Major Gaillard's observations appear to furnish conclusive evidence of the reliability of the trochoidal theory as regards its most important feature, the relation between length and speed of advance.

It is true that Major Gaillard dealt only with shallow-water waves, but it is evident from what has gone before that shallow-water trochoidal waves are more likely to misrepresent the actual waves than the deep-water trochoidal waves.

The actual wave profiles in the Duluth canal as obtained by photography agreed reasonably well with the profiles from the trochoidal formula. The differences, generally speaking, were greatest at about mid-height of the wave, where the failure of the trochoidal theory to satisfy the conditions of continuity and dynamical stability is most marked. Major Gaillard states that the elevated portion of an actual wave " is always narrower and the depressed portion broader and flatter than is indicated by theory, and this difference becomes more marked as the wave approaches the point of breaking." The actual wave profiles, however, were by no means uniform, differing from each other quite as much as from the trochoidal form.

To sum up it seems fair to say that the trochoidal formulæ represent actual waves very closely as regards speed, with a sufficient approximation as regards profile, and for practical purposes are much better than more complicated and difficult formulæ that have been devised. They are themselves quite complicated and difficult enough.

10. Shallow Water and Solitary Waves. — The trochoidal formula for wave speed in shallow water of depth d may be written

$$v^2 = \frac{e^{\frac{4\pi d}{L}} - 1}{e^{\frac{4\pi d}{L}} + 1} \cdot \frac{gL}{2\pi}.$$

For a constant length of wave v decreases as the water shoals, the ratio between the velocity of a wave of given length L in water of depth d below orbit centers and a wave of the same length in indefinitely deep water being

$$\sqrt{\frac{e^{\frac{4\pi d}{L}} - 1}{e^{\frac{4\pi d}{L}} + 1}}.$$

Fig. 28 shows a curve of the value of this ratio plotted on $\frac{d}{L}$. It is seen that for depths of water greater than half the wave length there is practically no change of speed.

Figs. 29 and 30 show graphically the relations between depth of water, length of wave and speed of wave, the speeds being expressed in knots per hour. Fig. 30 simply reproduces on a large scale for clearness the lower part of Fig. 29. It is seen that as the depth of water becomes very small the speed tends to become independent of the length. So let us investigate the results of assuming that the wave length is very much greater than the depth of water.

The formula for wave speed in shallow water is, as we have seen,

$$v^2 = \frac{b_0}{a_0}\frac{gL}{2\pi} = \frac{e^{\frac{4\pi d}{L}} - 1}{e^{\frac{4\pi d}{L}} + 1}\frac{gL}{2\pi}.$$

Now expanding we have

$$e^{\frac{4\pi d}{L}} = 1 + d\left(\frac{4\pi}{L}\right) + \frac{d^2}{2}\left(\frac{4\pi}{L}\right)^2 + \frac{d^3}{6}\left(\frac{4\pi}{L}\right)^3 + \cdots.$$

Then

$$v^2 = \frac{4\pi\frac{d}{L} + \frac{(4\pi)^2}{2}\left(\frac{d}{L}\right)^2 + \frac{(4\pi)^3}{6}\left(\frac{d}{L}\right)^3 + \cdots}{2 + 4\pi\frac{d}{L} + \frac{(4\pi)^2}{2}\left(\frac{d}{L}\right)^2 + \frac{(4\pi)^3}{6}\left(\frac{d}{L}\right)^3 + \cdots}\frac{gL}{2\pi}.$$

Now when $\frac{d}{L}$, or the ratio between depth and length, becomes very small all terms of the long fraction above except two can be neglected, and the fraction reduces to

$$\frac{4\pi\frac{d}{L}}{2} = 2\pi\frac{d}{L}.$$

Then

$$v^2 = 2\pi\frac{d}{L}\frac{gL}{2\pi} = gd.$$

In the above d is not the original depth of water but the depth to surface orbit centers, or to mid-height of the waves. This depth is somewhat greater than undisturbed still water depth, but not very much greater.

The above result is interesting as indicating that in shallow water, on the trochoidal theory, there is a limit to the speed of waves no matter what their length. This conclusion is confirmed by experience, and the value of the limit obtained above is in reasonable agreement with experiments. It is interesting to note in this connection that, as indicated in Fig. 25, the shoaler the water the more a trochoidal wave system tends to approach a series of sharp crests separated by long hollows that are nearly flat. That is to say, it tends to become a series of solitary waves, or waves of translation, consisting of humps or crests without hollows. Scott

Russell, as a result of numerous experiments on the so-called solitary wave, or wave of translation, made in a trough, concluded that the velocity of this wave was equal to that of a body falling freely through a height equal to half the depth from the top of the wave. The formula above gives the velocity of the trochoidal wave approaching the wave of translation type as that of a body falling through a height equal to half the depth measured from mid-height of the wave. The difference is not great for possible waves whose height is generally but a fraction of the depth. There is, however, testimony to indicate that Scott Russell's formula gives too great a velocity. Rankine gives a formula practically equivalent to Scott Russell's. Major Gaillard states that he has applied Rankine's formula to several hundred observations upon shallow-water waves, taken at North Beach, Fla., and on Lake Superior, and has found that it almost invariably gives results considerably in excess of the observed velocities. The trochoidal formula, then, with its velocity somewhat smaller than Scott Russell's or Rankine's, would agree more closely with Gaillard's observations.

11. Dimensions of Sea Waves. — It may be well to supplement the mathematical theory of waves with some information regarding waves found in practice. The heights of sea waves are their most striking feature and the most important for seagoing people. From the nature of the case it is very difficult to observe with accuracy the heights of deep-sea waves. From observations made by a number of observers of various nationalities in various seas it seems reasonable to consider that waves 40 feet high from trough to crest can be generated in deep water by unusually severe and long continued storms. This exceptional height is liable to be materially surpassed by abnormal waves, the result of superposition. Thus Major Gaillard quotes a case where a photograph taken on the United States Fish Commission steamer *Albatross*, and furnished him by Commander Tanner, U. S. N., showed the fore yard of the ship parallel to the crest of a huge wave and a little below it, the photograph being taken from aft. From the known dimensions of the vessel and position of the camera it seems that this crest must have been from 55 to 60 feet above its trough. This wave was photographed in the North Pacific off the United

States coast. Estimated heights as great as this are not infrequently reported by captains of steamers crossing the Atlantic, but accurate estimates of wave heights are difficult to make. Probably it would be a fair statement of the case to say that very heavy seas with maximum wave heights of 30 feet are not unusual. Exceptionally heavy seas with maximum wave heights of 40 feet are encountered at times, and there is good evidence that abnormal crests 60 feet in height have been encountered. The maximum wave height would not be found for every wave of a heavy sea. The 30 and 40 foot waves would appear at intervals. Intervening waves would be lower.

For the purpose of estimating the maximum stress of a ship it is customary to assume a wave height one-twentieth the length, the length of wave being taken the same as the length of the ship. This seems a reasonable average, but steeper waves have been often observed. Short waves are more apt to be steep than long waves. As to actual lengths it may be confidently stated that waves over 500 feet long are unusual, though a 40-foot sea would probably be between 600 and 800 feet long, and lengths of 1000 feet and more have been measured.

For the development of maximum waves a great space of open water is essential. Major Gaillard concluded after investigation that "during unusually severe storms upon Lake Superior, which occur only at intervals of several years, waves may be encountered in deep water of a height of from 20 to 25 feet and a length of 275 to 325 feet." It appears, then, that the 500-foot vessels navigating Lake Superior will probably never encounter waves their own length. This condition indeed is rapidly being reached by the enormously long Atlantic liners of the present day.

12. Relations between Wind and Waves. — The length of waves (or their speed of advance) is governed by the velocity of the wind creating the wave. The relation is not known. Waves have often been observed in advance of a storm and also waves in a storm that were traveling faster than the wind was blowing. It does not follow that a wave can travel faster than the wind that forms it. Severe storms are revolving or cyclonic, and the storm center does not move as fast as the wind blows. Hence a wave, though traveling more slowly than the wind that formed it, may run entirely ahead of the storm or into a region where the wind is blowing less violently.

Published observations upon the ratio of wave and wind velocity are not very concordant. Lieutenant Paris, of the French Navy, maker of very extensive and careful wave observations, gives the wave velocity as .6 that of the wind in a very heavy sea, and relatively greater as the sea becomes less heavy. Major Gaillard found at Duluth for waves in shallow water, which probably did not travel so fast as in the open lake, that the wave velocity as averaged from observations taken during fourteen storms was but .5 that of the wind. It appears probable that in a strong gale making a heavy sea the wave velocity is from .5 to .6 that of the wind, but that waves formed under these conditions often travel to regions where the wind is not blowing so fast as the waves are traveling.

If we take the wave formed as moving with .5 the speed of the wind we have from the trochoidal formula for deep water the following relations:

Speed of wind, statute miles	20	40	60	80	100
Speed of wave, statute miles	10	20	30	40	50
Speed of wave, f.s.	14⅔	20⅓	44	58⅔	73⅓
Length of wave, crest to crest, feet	42	168	378	673	1051

It would seem, then, if the above ratio between speed of wind and speed of wave is approximately correct, that waves more than 1000 feet in length should be very rare. As a matter of fact, they are very rare.

The height of storm waves will evidently depend upon the violence of the wind and the "fetch" or length of open water over which the wind blows. Mr. Thomas Stevenson, the noted British lighthouse engineer, established from many observations the following empirical formula:

$$h = c\sqrt{f},$$

where h is the wave height in feet, c is a coefficient depending upon the force of the wind, and f is the "fetch" in nautical miles. For strong gales the value of c is 1.5.

From this formula we have the following:

$h =$	10	15	20	25	30	35	40
$f =$	44	100	178	278	400	544	711

At first sight these results might appear inconsistent with the fact that waves more than 40 feet high are very rare, even where there are several thousand miles of open water. As a matter of fact, however, violent gales are revolving storms, and the violent part of such storms is seldom more than five or six hundred miles in diameter, so that Stevenson's formula is consistent with the general facts.

3. The Law of Comparison

1. Principle of Similitude. — Modern ideas of the resistance of ships are based largely upon the Law of Comparison, or Froude's Law, as it is generally called, connecting the resistance of similar vessels. By judicious application of this law we are enabled to determine, with fair accuracy, the resistance of a full-sized ship from the experimentally determined resistance of a small model of the same.

Froude's Law is a particular case of the general law of mechanical similitude, defining the necessary and sufficient conditions that two systems or aggregations of particles that are initially geometrically similar should continue to be at corresponding times not only geometrically but mechanically similar. The principle of similitude was first enunciated by Newton, but the demonstration now generally accepted we owe to French mathematicians of the last century. Mr. William Froude appears, however, to have developed independently the particular form used to compare ships and models and to have been the first to use the Law of Comparison to obtain useful practical results.

2. Deduction of Law of Comparison. — Suppose we have a particle of a system whose coördinates referred to rectangular axes are x, y and z. Let m denote the mass of the particle. If the particle is moving, it will have at time t an acceleration $\frac{d^2x}{dt^2}$ parallel to the axis of x, an acceleration $\frac{d^2y}{dt^2}$ parallel to the axis of y and similarly $\frac{d^2z}{dt^2}$ parallel to the axis of z. Let the components parallel to x, y and z of the external moving force upon the particle be denoted by X, Y and Z. Denote by δx, δy and δz the resolved motions parallel to the axes due to a small motion of the particle along its path.

Then using the well-known principle of Virtual Velocities, the differential equation giving the motion of the particle is

$$\left(X - m\frac{d^2x}{dt^2}\right)\delta x + \left(Y - m\frac{d^2y}{dt^2}\right)\delta y + \left(Z - m\frac{d^2z}{dt^2}\right)\delta z = 0.$$

Suppose, now, we have in a second system, mechanically similar, a corresponding particle of mass m' whose coördinates at time t', corresponding to time t in the first system, are x', y', z' and whose impressed force components are X', Y', Z'. Its equation of motion will be

$$\left(X' - m'\frac{d^2x'}{dt'^2}\right)\delta x' + \left(Y' - m'\frac{d^2y'}{dt'^2}\right)\delta y' + \left(Z' - m'\frac{d^2z'}{dt'^2}\right)\delta z' = 0.$$

If the motions of these two particles are geometrically and mechanically similar, the equations of motion must be the same, differing only by a constant factor. Now, for similar geometrical motions we have a constant ratio between x and x', etc.

Suppose $\quad x' = \lambda x, \qquad y' = \lambda y, \qquad z' = \lambda z.$

Then $d^2x' = \lambda d^2x$ and so on.

Let $m' = \mu m$, μ being the constant ratio of masses of the two particles.

Let the corresponding times be in the ratio T or $t' = Tt$ and $(dt')^2 = T^2 dt^2$.

Substituting for x', etc., their values we have

$$\left(X' - \mu m\frac{\lambda}{T^2}\frac{d^2x}{dt^2}\right)\lambda\delta x + \left(Y' - \mu m\frac{\lambda}{T^2}\frac{d^2y}{dt^2}\right)\lambda\delta y + \left(Z' - \mu m\frac{\lambda}{T^2}\frac{d^2z}{dt^2}\right)\lambda\delta z = 0.$$

This may be rewritten

$$\left(\frac{T^2 X'}{\mu\lambda} - m\frac{d^2x}{dt^2}\right)\delta x + \left(\frac{T^2 Y'}{\mu\lambda} - m\frac{d^2y}{dt^2}\right)\delta y + \left(\frac{T^2 Z'}{\mu\lambda} - m\frac{d^2z}{dt^2}\right)\delta z = 0.$$

Evidently, in order that this may become identical with the equation for the first system, we must have

$$\frac{T^2 X'}{\mu \lambda} = X \quad \text{or} \quad \frac{X'}{X} = \frac{\mu \lambda}{T^2}$$

and similarly

$$\frac{Y'}{Y} = \frac{\mu \lambda}{T^2} = \frac{Z'}{Z}.$$

It follows, then, that the external forces on corresponding particles must bear a constant ratio to each other. Let F denote this ratio. Then the necessary and sufficient relation for geometrical and mechanical similitude of motion of the two particles is $F = \frac{\mu \lambda}{T^2}$.

The same relation connects every corresponding particle of the two systems, and hence the systems as a whole. Now T, the relation ratio between corresponding times, is not very convenient for use in practical application. It is readily eliminated. Let v and v' be corresponding velocities. Then

$$v = \frac{dx}{dt}, \quad v' = \frac{dx'}{dt'} = \frac{\lambda}{T} \frac{dx}{dt}.$$

Whence $\frac{v'}{v} = \frac{\lambda}{T} = c$ say. Then $T^2 = \frac{\lambda^2}{c^2}$.

Whence $F = \frac{\mu \lambda c^2}{\lambda^2} = \frac{\mu c^2}{\lambda}$.

We may further simplify the case by assuming a relation between c and λ. Suppose we make the ratio of corresponding speeds such that $c^2 = \lambda$ or that the speed ratio is equal to the square root of the dimension ratio. Then $F = \mu$. Now we know that whatever the speed ratio and dimension ratio, the external forces due to gravity must be in the ratio μ or the ratio of masses. We see from the above that for motions mechanically and geometrically similar, if the speed ratio is made equal to the square root of the dimension ratio, all external forces must be in the ratio of mass or weight. The application to the case of a ship and its model is obvious. If a certain portion of the resistance of a ship is due to a certain disturbance of the water and if, at a corresponding speed of the model, bearing to the speed of the ship a ratio equal to the square root of the dimension ratio between model and ship, there is a similar disturbance set up by the model, the resistances due to the similar disturbances will be proportional to the weights of ship and model.

For the resistance of the ship or model, as the case may be, is in each case the external force, other than gravity, acting upon the system of particles involved in the disturbance, and the mass of disturbed water, if the disturbances are similar, is proportional to the displacement of the ship.

It is apparent from the above that the applicability of Froude's Law to resistances of model and ship depends upon whether the disturbances at corresponding speeds are similar. This is a matter capable of reasonably close experimental determination as regards the wave disturbances of model and ship. It is found that these are similar at corresponding speeds, the wave disturbance set up by the ship being an enlargement to scale as closely as can be measured of that of the model at corresponding speed.

Mr. William Froude estimated the actual resistance of the *Greyhound*, a ship of over 1,000 tons displacement, by applying the Law of Comparison to carefully measured resistances of a small model in a manner to be explained later, and found the results thus obtained in very close agreement with the actual resistance as measured by towing experiments. But, perhaps, the strongest experimental confirmation of the Law of Comparison, and one fully warranting its practical application, is an indirect one. There are now a number of experimental model basins in existence engaged in estimating the resistances of ships by proper application of the Law of Comparison to results of model experiments. These are not able to verify their results directly, because, for the full-sized ship when tried, we ascertain not resistance but the indicated power. The efficiency of propulsion connects the indicated power with the resistance. But, using the actual indicated powers and the estimated resistances determined from model results by the Law of Comparison, there are obtained efficiencies of propulsion which are consistent and reliable as a basis for new designs of vessels.

We are fully warranted, then, by numerous considerations,

both theoretical and practical, in reposing especial trust and confidence in Froude's Law. The modern theory of ships' resistance is founded upon it, and since it has been understood and utilized the numerous crude and treacherous theories which preceded Froude have practically disappeared.

It is possible to make a less general demonstration than the above of Froude's Law from the steady motion formula for stream lines. This, too, depends upon the similarity of stream lines around model and ship, a fact requiring experimental determination.

3. Applications of Law of Comparison. — Let us now determine the formulæ, etc., needed in the application of the Law of Comparison to ships' resistance.

Put into symbols, let L, B, H denote the length, breadth and mean draft of a ship in feet, D its displacement in tons and V its speed in knots. Let l, b, h, d, v denote similar quantities for a model of the ship. Suppose R and r denote resistances following Froude's Law. If λ denote the ratio between linear dimensions so that $L = \lambda l$, $B = \lambda b$ and so on and if V and v are connected by the relation $V = v\sqrt{\lambda}$,

then
$$\frac{R}{D} = \frac{r}{d}.$$

Since
$$\frac{D}{d} = \left(\frac{L}{l}\right)^3 = \lambda^3$$

we may write $R = \left(\frac{L}{l}\right)^3 r = \lambda^3 r$. It is to be noted, too, that

$$\lambda = \left(\frac{D}{d}\right)^{\frac{1}{3}} \quad \text{so} \quad V = v\left(\frac{D}{d}\right)^{\frac{1}{6}}.$$

The Law of Comparison is useful and applicable in connection with many problems besides that of the resistance of ships. Thus, it is directly applicable in comparing full-sized machines and their models of the same material. Here, too, since gravity is one external force always present, the speeds of corresponding parts must be in the ratio of the square roots of the linear dimensions. Thus consider a small and a large steam engine, similar and working at corresponding speeds. Let us find from the Law of Comparison the relations connecting pressures, revolutions, etc. Let R, T,

I, S and P denote, respectively, revolutions per minute, torque, indicated power, piston speed, and steam pressure for the large engine, and r, t, i, s, p the same quantities for the similar small engine or model. Let λ denote the ratio of linear dimensions. Then since the speeds must correspond, we have $S = s\sqrt{\lambda}$.

Now $\quad S = $ stroke of large engine $\times 2R$,

$\quad\quad\quad s = $ stroke of small engine $\times 2r$.

Also stroke of large engine $= \lambda$ stroke of small engine. Whence dividing $\dfrac{S}{s} = \lambda \dfrac{R}{r}$. But $\dfrac{S}{s}$ also $= \sqrt{\lambda}$. Whence $R = \dfrac{r}{\sqrt{\lambda}}$.

The total steam pressures on the pistons being the external forces must be in proportion to λ^3 and the piston areas are proportional to λ^2. Hence $P = \lambda p$. The indicated horse-power is proportional to the piston area, varying as λ^2, the steam pressure varying as λ and the piston speed varying as $\sqrt{\lambda}$. Hence on combining these three factors we have $I = i\lambda^{3.5}$. Now I is proportional to TR. Hence the torque is directly proportional to the indicated power varying as $\lambda^{3.5}$ and inversely proportional to the revolutions varying as $\dfrac{1}{\lambda^{\frac{1}{2}}}$.

Hence $T = t\lambda^{3.5} \div \dfrac{1}{\lambda^{\frac{1}{2}}} = \lambda^4 t$.

The above relations apply directly to centrifugal fans. For steam pressure we substitute the pressure at which the air is delivered. Also the quantity of air delivered will vary directly as the area of outlet pipe or as λ^2 and directly as the speed or velocity or as $\lambda^{\frac{1}{2}}$, whence at corresponding speeds the quantities of air delivered will vary as $\lambda^{2.5}$.

The above relations for revolutions, torque, power and pressure apply too to the operation of propellers. It should be noted since $P = \lambda p$ that the pressure per square inch of the water in which a propeller works should be λ times that of the water in which its model works. Model propellers are usually tested under a total head of 35 feet or so of water (equivalent to atmospheric pressure + one foot or so submersion below surface, say, 35 feet in all). For the pressure to vary linearly would require a full-sized propeller ten times as large as the model to work under a total head of 350 feet, or, say, 316 feet submersion, if the 34 feet head due to air pressure were equivalent in all respects to 34 feet of water. While this is only approximately the case, it is evident that the pressure conditions for model and propeller are not those required by the Law of Comparison. But it does not necessarily follow that the Law of Comparison would not apply to the conditions of practical operation. If the action of propellers is such that the power, torque and efficiency are unaffected by depth of submersion, the Law of Comparison would apply fully.

We shall see later that, under some conditions of operation, propeller action is but little affected by depth of submersion, while under others it is materially affected. Hence under some conditions the Law of Comparison applied to model propeller experiments may be expected to be a reliable guide, while under other conditions of operation it would certainly be fallacious.

Valuable and even indispensable as the Law of Comparison is in dealing with resistance and propulsion of ships, it must be applied with discretion and an understanding of its limitations. Some of these limitations will be developed later.

4. Simple Resistances Following Law of Comparison. — In reducing any kind of resistance to rule the endeavor is usually made to express it by a formula involving some power of the speed as V^2 or V^3. Unfortunately actual resistances of ships do not lend themselves to such simple formulæ, but it seems worth while to determine how resistances which satisfy the Law of Comparison and vary as definite powers of speed vary with displacement or dimensions.

Suppose $R = \phi(D) V^n$ expresses the law of variation of a ship resistance which satisfies the Law of Comparison, R being resistance in pounds, $\phi(D)$ some function of displacement, V the speed in knots and n an index according to which resistance varies.

For the similar models' resistance we have

$$r = \phi(d) v^n.$$

For corresponding speeds $\dfrac{V}{v} = \left(\dfrac{D}{d}\right)^{\frac{1}{6}}$ and $\dfrac{R}{r} = \dfrac{D}{d}.$

Then
$$\frac{R}{r} = \frac{\phi(D)}{\phi(d)}\left(\frac{V}{v}\right)^n = \frac{D}{d} = \frac{\phi(D)}{\phi(d)}\left(\frac{D}{d}\right)^{\frac{n}{6}}.$$

Whence $\phi(d)\dfrac{D^{\frac{n}{6}}}{D} = \phi(d)\dfrac{d^{\frac{n}{6}}}{d} = $ a constant regardless of displacement
$= C$ say.

Then
$$\phi(D) = CD^{1-\frac{n}{6}}$$

or
$$R = CV^n D^{1-\frac{n}{6}}.$$

For integral values of n we have the following results

$n = 1$ resistance varies as (displacement)$^{\frac{5}{18}}$ or (linear dimensions)$^{2\frac{1}{2}}$.
$n = 2$ resistance varies as (displacement)$^{\frac{2}{3}}$ or (linear dimensions)^2.
$n = 3$ resistance varies as (displacement)$^{\frac{1}{2}}$ or (linear dimensions)$^{1\frac{1}{2}}$.
$n = 4$ resistance varies as (displacement)$^{\frac{1}{3}}$ or (linear dimensions)^1.
$n = 5$ resistance varies as (displacement)$^{\frac{1}{6}}$ or (linear dimensions)$^{\frac{1}{2}}$.
$n = 6$ resistance is independent of displacement or dimensions.

The above results are not of much practical value since actual resistances even when following the Law of Comparison do not vary as simple powers of the speed, but they are of some use in connection with approximate formulæ.

5. Dimensional Formulæ. — In connection with the Law of Comparison it is of interest to note the so-called dimensional formulæ which are the functions of certain primary variables or units to which are proportional a number of things which we shall have occasion to use. Thus taking length or a linear dimension as a primary variable we have area varying for similar surfaces as (linear dimensions)^2 and similarly volume varies as (linear dimensions)^3. Then if we denote length or linear dimension by l we have l^2 and l^3 as the dimensional formulæ for area and volume respectively.

Similarly, if t denote time, since velocity varies directly as the length traversed in a given time and inversely as the time required to traverse a given length and is dependent upon no other variables, we have $\frac{l}{t}$ as the dimensional formula for velocity. Further, since acceleration varies inversely as the time required to gain velocity we have $\frac{l}{t^2}$ as the dimensional formula for acceleration.

The practical application of dimensional formulæ is mostly in connection with conversion factors for the determination of the numerical magnitude or numbers representing definite things when the fundamental units are changed. Thus, suppose we have a length of 24 feet. If the yard were the unit of length this length would be expressed numerically by 8 instead of 24. Similarly, suppose we have a surface of 108 square feet. If the yard were the primary unit the number of units of surface would be $\frac{108}{(3)^2} = 12$.

Since the dimensional factor for area is l^2 the conversion factor is the square of the ratio of the linear units. Similarly the conversion factor for volume is the cube of the ratio of linear units and 135 cubic feet would be $\frac{135}{(3)^3} = 5$ cubic yards. These transformations are puzzling in some cases and it will be well to give the general rule applicable.

We will have in any given case the old number, or the number expressing something quantitatively in the old units, the ratios between the units or the numbers expressing the new units in the old units and vice versa, and the dimensional formula for the thing under consideration — area, volume, velocity or what not.

Then express the old unit of each kind in terms of the new and substitute in the dimensional formula for each primary variable the corresponding numerical ratio $\frac{\text{old unit}}{\text{new unit}}$. The result is the conversion factor, and we have

$$\text{New number} = \text{Old number} \times \text{Conversion factor.}$$

Thus when converting square feet to square yards the ratio $\frac{\text{old length unit}}{\text{new length unit}} = \frac{1}{3}$. The dimensional formula is l^2. Then

$$\text{Conversion factor} = \left(\frac{1}{3}\right)^2 = \frac{1}{9}.$$

Old number = 108.

$$\text{New number} = 108 \times \frac{1}{9} = 12.$$

Similarly, suppose we have a velocity of 69.3 feet per second and wish to convert it into statute miles per hour.

For velocity the dimensional formula is $\frac{l}{t}$.

$$\frac{\text{Old length unit}}{\text{New length unit}} = \frac{1}{5280} \qquad \frac{\text{Old time unit}}{\text{New time unit}} = \frac{1}{3600}.$$

$$\text{Conversion factor} = \frac{1}{5280} \div \frac{1}{3600} = \frac{3600}{5280} = \frac{15}{22}.$$

$$\text{New number} = 69.3 \times \frac{15}{22} = 47.25 \text{ statute miles per hour.}$$

By following the above method strictly and systematically there is no difficulty in obtaining correct conversion factors no matter how complicated the dimensional formulæ.

It is usual to use as primary variables in dimensional formulæ for things with which we are concerned length denoted by l, time denoted by t, and mass denoted by m.

Since, however, velocity, denoted by v, is proportional to $\frac{l}{t}$ or t is proportional to $\frac{l}{v}$, we may use m, l and v as primary variables.

Further, if, as in the Law of Comparison, we assume certain relations to exist between l and m and l and v, we can express dimensional formulæ in terms of l alone. For the Law of Comparison we assume m to vary as l^3 and v to vary as $l^{\frac{1}{2}}$. The table below gives the dimensional formulæ of importance for our purposes.

	Dimensional Formulæ.		
	In $m, l, t.$	In $m, l, v.$	In l alone when Law of Comparison relations between m, l and v hold
Length..........................	l	l	l
Area or surface...................	l^2	l^2	l^2
Volume........................	l^3	l^3	l^3
Angular velocity and revolutions per minute	$\dfrac{1}{t}$	$\dfrac{v}{l}$	$\dfrac{1}{l^{\frac{1}{2}}}$
Angular acceleration	$\dfrac{1}{t^2}$	$\dfrac{v^2}{l^2}$	$\dfrac{1}{l}$
Linear velocity...................	$\dfrac{l}{t}$	v	$\sqrt{l}$
Linear acceleration................	$\dfrac{l}{t^2}$	$\dfrac{v^2}{l}$	1
Density........................	$\dfrac{m}{l^3}$	$\dfrac{m}{l^3}$	1
Moment of inertia.................	ml^2	ml^2	l^5
Momentum......................	$\dfrac{ml}{t}$	mv	$l^{3\frac{1}{2}}$
Moment of momentum or angular momentum	$\dfrac{ml^2}{t}$	mvl	$l^{4\frac{1}{2}}$
Force or resistance	$\dfrac{ml}{t^2}$	$\dfrac{mv^2}{l}$	l^3
Work, energy and torque...........	$\dfrac{ml^2}{t^2}$	mv^2	l^4
Power....	$\dfrac{ml^2}{t^3}$	$\dfrac{mv^3}{l}$	$l^{3\frac{1}{2}}$
Pressure or stress per unit area......	$\dfrac{m}{lt^2}$	$\dfrac{mv^2}{l^3}$	l

It will be observed that the relations in the third column agree with those deduced in various specific cases when considering the Law of Comparison.

4. Wetted Surface

1. Importance of Surface Resistance. — For all but a minute proportion of actual steam vessels the skin friction resistance, or the resistance due to friction of the water upon the immersed hull surface, is greater than the resistance due to all other sources of resistance combined. For some of the fastest Atlantic liners, for instance, the skin resistance at top speed, under ordinary smooth-water conditions, is about 64 per cent of the total resistance. For only the comparatively few vessels that are pushed to a speed very high in proportion to their length does the residuary resistance due to all causes surpass the skin resistance.

Such extremely fast vessels are nearly all for naval purposes. They are seldom warranted by commercial conditions.

In view of the great importance of the Skin Resistance it is advisable to make a careful investigation into the question of the wetted surface of ships. We need to know how to calculate it accurately, and how to estimate it with close approximation. We need, too, if the question of wetted surface is to be given its proper influence in design work, to understand the relations between wetted surface and size, proportions and shape of ships.

2. **Appendage Surface.** — The wetted surface of hull appendages can be calculated as a rule without difficulty. Appendages of importance have nearly always plane or nearly plane surfaces, and their areas are readily determined by straightforward processes. Appendage surface, then, can be calculated by simple methods, the exact procedure varying with circumstances. In dealing with such appendages as bilge keels and docking keels, which cover or mask some of the surface of the hull proper, it is best to deduct from the gross area of the appendage the area masked by it, the net area resulting being the addition to the wetted surface of the hull proper due to the presence of the appendage.

3. **Surface of Hull Proper.** — When we undertake the accurate calculation of the wetted surface of the hull proper of a ship, we encounter at once a serious difficulty. It is not possible to develop or unroll into a single plane the curved surface of a ship's bottom. We can draw a section at any point and measure its girth, and if the ribbon of surface included between two sections a foot apart were equal in area to the girth in feet of the section in the middle of the ribbon, it would be very simple to determine accurately the wetted surface of the hull proper by applying Simpson's Rules or other integrating rules of mensuration to a series of girths at equi-

distant stations, covering the whole length of the ship. Unfortunately, however, on account of its obliquity, the area of this ribbon of surface is in general appreciably greater than its mid girth, and for the best results we must devise a more accurate method. The simplest plan is to correct the mean girth in question, multiplying it by a suitable factor, so that the area of the ribbon will be equal to the corrected mid girth. Then we can apply the ordinary rules to the corrected mid girth and obtain accurate results. Let us see now how to determine the correction factor — first for one point of a section and then for a whole section.

4. **Obliquity Factors.** — In Fig. 31 suppose AB, drawn straight for convenience, to represent a short portion of a section of a ship's surface by a normal diagonal plane. CD is parallel to the fore and aft line. Let AB cut the section FE in E and adjacent parallel sections each six inches from FE at L and K. Fig. 32 shows diagrammatically the three sections and the diagonal plane on the body plan. The oblique line KL is an element of surface, and we want to connect its length with ML, the distance between stations. Now, $KL = ML \sec KLM$. Hence $\sec KLM$ is the factor we need. Now $\tan KLM = \dfrac{KM}{LM} = \dfrac{mk \text{ in Fig. 32}}{LM}$. In practice, then, if we take a point on a section midway between two other or end sections, draw a line on the body plan at the point perpendicular to the section and measure the intercept (mk in Fig. 32) between the two end sections, we have

Tangent of angle of obliquity $= \dfrac{mk}{\text{distance between end sections}}$

and correction factor for obliquity at the point = secant of angle of obliquity.

We do not want to calculate tangents and secants, and we wish to work directly from the body plan. So we divide the sections on the body plan at six points into five equal parts. The most satisfactory method is to lay off small chords with a pair of dividers and thus determine the points of division by trial and error. Then we prepare a paper scale so divided that when set perpendicular to a section at a division point we read at once the correction factor for obliquity from the intercept between the two sections adjacent to the one for which we are determining obliquity factors. The paper scale can be laid off graphically, but can also be readily calculated. Let us suppose that the actual distance apart of successive sections in the sheer or half-breadth plan is 1 inch. Then the distance between two sections on either side of a middle section will be 2 inches. Suppose at a certain point the intercept of the perpendicular in the body plan between the two stations adjacent to the one we are considering is 0.25 inch. Then the tangent of the angle of obliquity is $\frac{0.25}{2} = .125$. Hence at this point the angle of obliquity is $7° 7'\frac{1}{2}$ since $\tan^{-1} .125 = 7° 7'\frac{1}{2}$. The correction factor at the point is sec $7° 7'\frac{1}{2}$ or 1.00778. Then for our scale $\frac{1}{4}$ inch corresponds to a correction factor of 1.00778. But to lay off our scale we want to determine the varying lengths corresponding to equal intervals of correction factor.

The necessary calculations are shown in Table II which applies directly to 1-inch section spacing.

Of course, an actual set of lines would nearly always have sections spaced more than 1 inch on the plans. For instance, a ship 416 feet long between extreme stations, with 21 stations or 20 spaces, would, if the plans were on the scale of $\frac{1}{4}$ inch to the foot, have the sections on the plans spaced $\frac{416}{20} \times \frac{1}{4} = 5.2$ inches. For such a ship the data for laying off the proper obliquity scale would be obtained by multiplying the figures in Column 4 of Table II by 5.2.

5. Sample Calculations. — Fig. 33 shows an actual body plan with each section divided into five equal spaces for the purpose of measuring obliquity and an obliquity scale in place measuring a correction factor of 1.015 for a point on section No. 15. Table I shows the calculations in standard form. It is seen that for each section the average correction factor for obliquity is calculated from the measurements at six points. The actual measured mean girths having been corrected, the wetted surface is readily calculated. The trapezoidal rule is used for the work, being really as accurate as Simpson's for curves of the type to be handled, and much shorter.

6. Average Correction Factors. — It is seen that the correction factors for obliquity are always very close to unity. Advantage may be taken of this fact when dealing with ships of ordinary form to utilize average correction factors which, when multiplied into the product of the mean girth by the length, will give the wetted surface with great accuracy, i.e., within a small fraction of one per cent.

Fig. 34 gives contour curves of correction factors for obliquity plotted upon values of $\frac{L}{B}$, or ratio between length and beam, and $\frac{L}{H}$, or ratio between length and draught.

For vessels of ordinary form it will be found that by determining the mean girth and applying the correction factor from Fig. 34 the wetted surface is determined with substantially the same accuracy as if complete calculations had been made. Fig. 34 must be used with caution for vessels not of ordinary form, if very accurate results are wanted.

7. Girths of Sections. — Having seen how to determine with accuracy the wetted surface of a ship of which complete plans are available, I will now take up the determination of the approximate wetted surface of a vessel whose dimensions and displacement are known, but for which complete plans are not yet available. This is a calculation which must often be made. Consider first the question of the girth of a ship section below water. This varies with dimensions, proportions, and shape or fullness of section. The variation with dimensions is a very simple matter. For similar sections the girth varies as any linear dimension, such as beam, or draught or $\sqrt{\text{area}}$. It is convenient to use $\sqrt{\text{area}}$ as governing quantity and express the girth G of a section of area in square feet $= A$ by $G = g\sqrt{A}$. For all similar sections of varying dimensions the quantity g in the formula preceding is constant. It is in fact the girth of a section of one square foot area and similar in all respects to the section whose area is A. Being a measure as it were of the girth, let it be called the girth parameter. We want now to ascertain how the girth parameter of a section varies with proportions and shape. The girth parameters of a few simple sections are obvious. Thus, if we have a square section of one square foot area, the beam is equal to the draught and the girth is 3 feet, or the girth parameter is 3. If the section is rectangular of $\frac{1}{2}$ foot draught and 2 feet beam, the girth parameter is again 3. We can in fact express by a formula the girth parameter of a rectangular section of any proportions. Let B denote its beam, xB its draught. Then xB^2 is its area A, and $B + 2xB$ the girth G. Now the girth parameter $g = \dfrac{G}{\sqrt{A}} = \dfrac{B + 2xB}{\sqrt{B^2 x}} = \dfrac{1 + 2x}{\sqrt{x}}$.

Fig. 35 shows a curve of girth parameter for rectangular sections plotted on x. The minimum value is 2.8284 for $x = \frac{1}{2}$, for which, if the section is one square foot in area, the beam is 1.4142 and the draught is .7071. For a semicircle of radius r the area $= \dfrac{\pi r^2}{2}$ and the girth πr. Whence $g = \dfrac{\pi r}{\sqrt{\dfrac{\pi r^2}{2}}} = \sqrt{2\pi} = 2.5066$. This value 2.5066 for a semicircle appears to be the minimum girth parameter possible. The sectional coefficient for a circle is .7854, and, as will be seen, this coefficient is close to that for a minimum girth for any proportion of beam and draught.

8. Actual Girth Parameters. — The best way to investigate the variation of girth parameter with proportions and fullness of section is to draw a number of sections of varying proportions and fullness and determine and plot their girth parameters. This has been done for a large number of sections covering a wide range of fullness and proportions. These sections were all calculated from the same basic formula, the variations of fullness, etc., being obtained by variation of coefficients. The details of the work are somewhat voluminous and need not be given. The results are fully summarized in Fig. 36, which gives contour curves of girth parameter plotted upon values of $\dfrac{B}{H}$ and sectional coefficient.

Fig. 36 is not, of course, applicable to freak or abnormal sections, but throughout its range is believed to be practically exact for sections of usual type.

For instance, Fig. 37 shows a series of sections of which No. 1 is a parabola and No. 6 is made up of two straight lines and the quadrant of a circle. The other four sections divide into five equal parts the intercepts between 1 and 6 of diagonal lines through O. Four other figures similar to Fig. 37, except that they had different proportions, were drawn, and the areas and girth parameters of the 30 sections thus obtained were carefully determined. Table III shows these actual girth parameters and girth parameters for the same proportions and fullness as taken from Fig. 36. The actual girth parameters were calculated to the nearest figure in the third place only.

It is seen that Fig. 36 applies to the curves of Fig. 37 and the other derived figures with great accuracy.

As instancing its application to actual ships' sections attention is invited to Table IV. This gives for 20 actual midship sections of vessels whose dimensions and proportions are stated, the actual girth parameters as measured and the girth parameters from Fig. 36 for sections of the same proportions and coefficients. The agreement is very close indeed.

It is evident from Tables III and IV that Fig. 36 represents with great accuracy the variation of girth parameters of usual sections of ships as dependent upon ratio of beam to draught and coefficient of fullness. It follows that, substantially, these are the only variables. That is to say, if we settle the beam, draught and area of a section of usual type, we substantially settle the girth, which varies but little with possible changes of shape. Of course, this does not apply to sections that are very hollow, having coefficients well below .5. Fig. 36 does not cover such sections, nor sections of extreme proportions of draught to beam, such as forward and after deadwoods. For such sections the girth parameters vary with great rapidity for small changes of beam. Fig. 36, however, covers nearly all the sections of actual ships of usual form and is worthy of careful study. We see from it that there is an actual minimum girth parameter a little greater than 2.5 occurring for $\frac{B}{H} = 2$ and coefficient of fullness a little below .8. Probably we may safely call the coefficient for minimum girth parameter .7854, the coefficient for a circle. Roughly speaking, as we vary $\frac{B}{H}$ the minimum girth parameter is always found for sectional coefficient in the neighborhood of .8 until we get to low values of $\frac{B}{H}$, below 1.5, where the minimum girth parameters correspond to larger coefficients. Similarly, as we vary sectional coefficient only the minimum girth parameter corresponds very closely to $\frac{B}{H} = 2$ until we reach coefficients greater than .9, when it corresponds to smaller values of $\frac{B}{H}$. The most striking feature of Fig. 36, however, is the comparatively small variation of girth parameter over a range of values of $\frac{B}{H}$ and sectional coefficient which covers the bulk of the sections of actual ships. This fact is of great importance in connection with the determination of a reliable approximate formula for wetted surface and the consideration of the influence of dimensions, proportions and shape upon wetted surface.

9. Approximate Formula for Wetted Surface. — Suppose we take $n + 1$ sections of a given ship, equally spaced at $n + 1$ stations 0, 1, 2, 3 . . . n. For each section, with subscript denoting the station, denote the girth by G, the girth parameter by g and the area by A. Let L denote the length and $\bar{G}$ the mean girth. Then

$G_0 = g_0 \sqrt{A_0}$, $G_1 = g_1 \sqrt{A_1}$ and so on.

Using the trapezoidal rule we have

$$\bar{G} = \frac{\tfrac{1}{2}G_0 + G_1 + G_2 \ldots G_{n-1} + \tfrac{1}{2}G_n}{n},$$

$$= \frac{\tfrac{1}{2}g_0\sqrt{A_0} + g_1\sqrt{A_1} + g_2\sqrt{A_2} + \cdots + g_{n-1}\sqrt{A_{n-1}} + \tfrac{1}{2}g_n\sqrt{A_n}}{n}.$$

Let S denote the wetted surface. Then neglecting obliquity, which will take care of itself later, when we determine coefficients from actual ships, we have

$$S = \bar{G}L = L\frac{\tfrac{1}{2}g_0\sqrt{A_0} + g_1\sqrt{A_1} + g_2\sqrt{A_2} + \cdots + g_{n-1}\sqrt{A_{n-1}} + \tfrac{1}{2}g_n\sqrt{A_n}}{n}.$$

If we keep the same sections and space them twice as far apart, we double length and displacement. We also, neglecting obliquity, double the wetted surface. If we keep length the same and double the area of each section, we double displacement. The girth parameters of the individual sections are unchanged, so that the result is to multiply S by $\sqrt{2}$. Now, what convenient expression involving only length and displacement will give us the same variation? Evidently, if we write $S = C\sqrt{DL}$ where D is displacement in tons, L is mean immersed length in feet and C is a coefficient depending upon proportions, shape, etc., but not upon dimensions, we have an expression for S which will vary for similar vessels just as the almost rigorous expression deduced above. For, if we double length and displacement, we double S; if we keep L constant and double D, we multiply S by $\sqrt{2}$.

As regards primary variation, then, this expression is as accurate as the rigorous one. It should be carefully noted that L in this formula is the mean immersed length, or the average water line length. In many types of vessels the water line lengths are sufficiently close to the mean immersed lengths to be used without error, but in others, the stem and stern profiles are such that for accurate work the mean immersed lengths must be determined. For rough work and first approximations before we are in a position to determine from plans the mean immersed length, load water line length is used. Secondary variation in the rigorous expression given above can come only with variations of the girth parameters, g_0, g_1, etc. The principal factors affecting the girth parameters are, as we have seen, variations of ratio of beam and draught and variations of sectional coefficient. Our formula $S = C\sqrt{DL}$ so far takes no direct account of these. They will show themselves in variations of the coefficient C from ship to ship.

10. Variation of Wetted Surface Coefficient. — Consider, first, the effect upon wetted surface coefficient of the ratio between beam and draught. This variation is most conveniently referred to the value of $\dfrac{B}{H}$ for the midship section. Fig. 38 shows the variation of wetted surface with the variation of $\dfrac{B}{H}$ for the lines of the

United States Practice Vessel *Bancroft*. Keeping length and displacement constant, a number of body plans were drawn from her lines with $\frac{B}{H}$ varying from 1 to 6. The wetted surface for each ratio was calculated and the resulting curve is shown plotted on $\frac{B}{H}$ in Fig. 38. It is seen that the minimum wetted surface is found at $\frac{B}{H} = 2.8$; but as $\frac{B}{H}$ is changed the variation is slow until we reach small values of $\frac{B}{H}$, when the wetted surface begins to increase rather rapidly. Such small values of $\frac{B}{H}$, by the way, are below values found in practice. The general features of Fig. 38 could be inferred from Fig. 36. We see from the latter figure that for a single section the minimum value of g is found for $\frac{B}{H} = 2$. Now, if for the midship section we had $\frac{B}{H} = 2$, the girth parameter of this one section would be a minimum, but for every other section the girth parameter would be above the minimum, since for every other section $\frac{B}{H}$ would be less than 2. Also for the smaller values of $\frac{B}{H}$ the girth parameters increase more rapidly than for the larger values. Hence, for actual ship lines of given length and displacement, but varying $\frac{B}{H}$, the minimum wetted surface must correspond to a value of $\frac{B}{H}$ greater than 2, and the wetted surface would increase, of course, on each side of the minimum. This minimum is found at $\frac{B}{H} = 2.8$ in Fig. 38.

It is not so easy to connect the variations of girth parameter of an actual ship with variations of sectional coefficient. Furthermore, Fig. 36 shows such small variation of girth parameter for sectional coefficients ranging from .7 to .9 that we may expect to find in practice the variation due to sectional coefficient masked by other arbitrary causes impossible to reduce to rule, such, for instance, as unusual amount of deadwood or extreme reduction of deadwood.

However, broadly speaking, the fuller the midship section, the fuller all the sections are likely to be, and, if the midship section is very fine, all sections are likely to be fine. These principles considered with Fig. 36 would lead us to expect in practice, when using the formula $S = C\sqrt{DL}$, to find rather high values of C associated with very fine midship sections, and possibly a minimum value of C for a fairly high midship section coefficient.

In this connection attention is invited to Figs. 39 and 40, which show variation of wetted surface coefficient with midship section coefficient, Fig. 39 for fine ended models and Fig. 40 for full ended models. The four curves in each figure refer to different values of the coefficient $D \div \left(\dfrac{L}{100}\right)^3$ as indicated. The higher values of wetted surface coefficient are found with the higher values of the coefficient $D \div \left(\dfrac{L}{100}\right)^3$. This is to be expected, since the greater the displacement on a given length the greater the obliquity. Figs. 39 and 40 refer to a single ratio of beam to draught, namely 2.923, but they show distinct minimum values of wetted surface coefficient in the neighborhood of midship section coefficients of .90. As regards absolute values of the coefficients it is to be noted that at midship section coefficient .84 they are practically coincident. For higher values of the midship section coefficient the fine ended models have the smaller wetted surface. For smaller values of midship section coefficient the fine ended models have the greater wetted surface. The extreme variations of coefficients in Figs. 39 and 40 are but about 3 per cent above and below the average, a fact which shows that the coefficient C in the approximate formula is nearly constant in practice.

11. Average Wetted Surface Coefficients. — Figs. 39 and 40 refer to models of only two types of lines.

A large number of actual wetted surfaces for many types of lines have been calculated at the model basin from which Fig. 41, showing contour curves of the wetted surface coefficient C plotted on $\dfrac{B}{H}$ and midship section coefficient, has been deduced.

The wetted surface coefficients of Fig. 41 were obtained from average results of vessels of ordinary form. For such vessels, if the mean immersed length is accurately known, they are correct within a small percentage. They apply to the hull proper only, exclusive of appendages, and should be used with caution for vessels of abnormal form, such as very shallow draught vessels, vessels with very broad, flat sterns, vessels with deadwood cut away to an unusual extent, etc.

In practice Fig. 41 can be utilized to ascertain with a good deal of accuracy the wetted surface of a vessel of abnormal type, provided we have the correct value of C for one vessel of the type which does not differ too much in proportions and coefficients from the vessel whose wetted surface is needed.

For, suppose that Fig. 41 is 4 per cent in error for the abnormal vessel whose wetted surface coefficient is known. It will continue to be very approximately 4 per cent in error for the type of lines under consideration as proportions and coefficients are changed, and its results corrected by 4 per cent may be relied upon for the abnormal type. In other words, Fig. 41 may be utilized in two ways:

a. To ascertain the approximate wetted surface of any vessel of ordinary type whose dimensions, displacement and midship section area only are known.

b. To ascertain the approximate wetted surface of a vessel of extraordinary type of known dimensions, etc., provided we know the actual wetted surface of another vessel of the same extraordinary type.

From a consideration of what has gone before, and especially of Figs. 36 to 41, we appear to be warranted in drawing a few broad conclusions as to the wetted surface of vessels of usual types.

1. For a given displacement the wetted surface varies mainly with length, being nearly as the square root of the length.
2. For a given displacement and length the wetted surface varies

but little within limits of beam and draught possible in practice. As regards wetted surface the most favorable ratio of beam to draught is a little below 3.

3. For given displacement and dimensions the wetted surface is affected very little by minor variations of shape, etc. Extremely full sections are somewhat, and extremely fine sections are quite prejudicial to small surface.

4. After length, the most powerful controllable factor affecting wetted surface is probably that of deadwood. By cutting away deadwood boldly, we can often save more wetted surface on a ship of given displacement and length than by any practicable variation in ratio of beam to draught, or in the fullness of sections.

5. Focal Diagrams

1. Field for Focal Diagrams. — In attempting to analyze experimental data it frequently happens that we know the general law which we think should govern, and we wish to examine whether the law does apply and, if it does, to determine suitable coefficients from the experiments for use in the formula expressing the law. Experimental data being at best an approximation, it is desirable to use a method which will not only give us an adequate approximation to the coefficients or constants desired, but give us some idea as to how closely our results are going to represent the observed data.

Mathematically, the problem is in general one of Least Squares. In practice, for many problems there is one coefficient or constant to be determined, the actual determination, of course, being made by taking average results. In a great many cases not so simple there are two coefficients or constants involved. For such cases, instead of applying the complicated and laborious methods of Least Squares, very satisfactory results can always be obtained from data not too much in error by the use of what I may call a Focal Diagram.

2. Illustration of Focal Diagrams. — This method may be readily comprehended from a concrete illustration. Fig. 42 shows a parabola whose equation is $y = 3x - \dfrac{x^2}{4}$, the general equation being of the form $y = ax - bx^2$. At the point P, say, where $x = 4$, $y = 8$. Substituting these values of x and y in the general equation, we have $8 = 4a - 16b$. This is a linear relation between a and b, and laying off axes of a and b as in Fig. 43, we can draw a line representing this relation. If we take the simultaneous values of a and b for any point on this line and substitute them in the general formula $y = ax - bx^2$, the resulting parabola in x and y would pass through $y = 8$, $x = 4$.

Fig. 43 shows ten lines in a and b corresponding to ten points on the parabola $y = 3x - \dfrac{x^2}{4}$. These points are as follows:

$x = 1 \quad 2 \quad 3 \quad 4 \quad 5 \quad 6 \quad 7 \quad 8 \quad 9 \quad 10$
$y = 2.75 \quad 5 \quad 6.75 \quad 8 \quad 8.75 \quad 9 \quad 8.75 \quad 8 \quad 6.75 \quad 5$

These ten lines all pass through the point $a = 3$, $b = .25$, forming a focus at this point. Evidently, if we know the x and y values of the ten points and the fact that they are on a curve whose formula is of the form $y = ax - bx^2$, we could determine a and b by drawing the ten lines as in Fig. 43 and taking the focal values $a = 3$, $b = .25$. If we knew the exact ordinates of but two spots, we could draw the two corresponding lines in Fig. 43 and determine the values of a and b.

In practice, if we determined the spots on the curve by experiment or observation, we would have more spots than theoretically needed to determine the focus; but the line for each spot instead of passing through the focus would pass somewhat near it, its distance from the focus depending upon the nearness of our observations to exact truth.

In Fig. 42, circles on the curve indicate ten exact spots, and adjacent crosses indicate spots of varying errors in location. The errors, both vertical and horizontal, vary by .05 from $+.25$ to $-.25$, and the actual errors at any spot were assigned by lot.

We have, then, for the approximate spots

$x = 1 \quad 1.75 \quad 2.85 \quad 4.10 \quad 4.80 \quad 5.90 \quad 7.20 \quad 8.05 \quad 9.25 \quad 9.95$
$y = 2.55 \quad 4.85 \quad 6.65 \quad 8.15 \quad 8.75 \quad 9.20 \quad 8.85 \quad 7.75 \quad 7.00 \quad 5.05$

A focal diagram similar to Fig. 43 can be drawn with a line for each approximate spot, and this is done in Fig. 44. It is evidently possible in Fig. 44 to spot the focus with an accuracy ample for most practical purposes.

3. Considerations Affecting Focal Diagrams. — If the assumed law or general equation is materially in error, a good focus will not be formed, no matter how close the observations may be. Even with an exact law it may be difficult to locate the focus if the observations are poor, but when we do get a good focus we know at once that the corresponding values of the coefficients in our formula will cause the formula to represent the experimental results with great accuracy, indicating that the assumed formula is close to the truth and that the observations are good.

In Fig. 44 the lines are straight. This need not necessarily be the case. The relation between a and b may not be linear, but can always be represented by a curve. Linear focal diagrams are, however, much the simplest and best and should always be sought for. Frequently, when the relation between the coefficients is not linear, it may be made so by adopting new coefficients of definite relation to the original ones.

In a linear focal diagram we usually determine two points on each line. The exact methods best to use vary somewhat with the nature of the case. It is always desirable to determine the two points, one on either side of the focus. Below are given the detailed calculations for the case we have been considering from the results of which Fig. 44 was plotted.

Formula: —

$$y = ax - bx^2, \quad a = \frac{y}{x} + bx, \quad b = 0, \quad a = \frac{y}{x}, \quad b = .5, \quad a = \frac{y}{x} + .5\,x.$$

x		1	1.75	2.85	4.10	4.80	5.90	7.20	8.05	9.25	9.95
y		2.55	4.85	6.65	8.15	8.75	9.20	8.85	7.75	7.00	5.05
$\frac{y}{x} = a$ for $b = 0$		2.550	2.771	2.333	1.988	1.823	1.559	1.229	0.963	0.757	0.508
$.5\,x$		.500	.875	1.425	2.050	2.400	2.950	3.600	4.025	4.625	4.975
a for $b = .5$		3.050	3.646	3.758	4.038	4.223	4.509	4.829	4.988	5.382	5.483

6. The Disturbance of the Water by a Ship

The disturbance of deep water by a ship passing through it is a very complicated matter and the disturbance of shallow water even more complicated. Broadly speaking, all resistance is due to disturbance of the water and before considering in detail the elements of resistance it will be well to form some idea of the nature of the disturbances to which resistance is due.

1. **Comparison between Ideal Stream Motion and Actual Motion.** — Suppose we could apply on the surface of the water a rigid frictionless sheet as of ice surrounding to a great distance a moving ship and advancing with it. If the ship had a smooth and frictionless bottom and the water were a perfect liquid there would be perfect stream line motion, and we know from stream line considerations the salient characteristics of what may be called the stream line disturbance in the vicinity of the ship's hull. In the vicinity of and forward of the bow the water would be given a forward and outward motion, with pressure in excess of that of the undisturbed water. Passing aft, the water would continue to flow outward, but at a short distance abaft the bow would lose its forward motion and begin to move aft as well as outward. Its pressure, a maximum near the bow, would steadily fall off, soon becoming less than that of undisturbed water.

Abreast the midship section, the sternward velocity would reach a maximum and the pressure a minimum. Passing sternward, as the water closed in it would lose its sternward velocity, and pressure would increase again until in the vicinity of the stern we would have excess pressure and the water would have motion forward as at the bow. Since there would be a deficiency of pressure over the greater portion of the hull, we must, in order to realize the ideal motion, assume that the rigid sheet surrounding the ship is strong enough to hold it firmly at the level at which it naturally floats when at rest. We must also assume that the pressure of the undisturbed water is such that the defect of pressure caused by the motion of the ship will not cause the water to fall away from the rigid sheet.

Now the motion of the actual ship through actual water differs from the ideal conditions assumed above.

1. The water is not frictionless, but is affected by the frictional drag of the surface of the ship.

2. The ship is not constrained to remain at a fixed level, but may rise and fall bodily and change trim in response to the reactions of the water.

3. The water surface is not constrained to remain at one level, but is free to rise and fall in response to the action of the ship.

2. Changes of Level of Vessel and Water. — Notwithstanding the differences between the actual circumstances of the motion and the ideal conditions assumed above, there is no doubt that the stream line action around an actual ship presents in a qualitative way nearly all the features of the ideal case considered. But in the actual ship the excess pressures at bow and stern result in surface disturbances, causing waves which spread away and absorb energy, and the defect of pressure amidships results in a lowering of the water level and a lowering of the ship bodily, accompanied by a change of trim.

Figs. 45 to 49 show for two speeds of one model and three speeds of another changes of level and trim of model and of level of water against the side. The dimensions and displacements of the models are given in the legend just above Fig. 45.

These figures are typical. They show elevations of the water at bow and stern, and show further two phenomena already described as to be expected from stream line action but not conspicuous or easy to determine for an actual ship. It is seen that there is a bodily settlement of the vessel and that in the vicinity of the mid length there is a bodily lowering of the water surface adjacent to the ship independent of the disturbance due to the wave created at the bow.

3. Lines of Flow over Surface of Vessel. — There have been a number of experiments made at the United States Model Basin upon the direction of relative flow of the water in the vicinity of models. The model surface being coated with sesquichloride of iron mixed with glue, pyrogallic acid is ejected at a point of the bottom through a small hole, which as it passes aft mingled with the water causes a gradually widening smear of ink upon the prepared model surface. The center line of this smear can be located with reasonable accuracy for some distance, and when it becomes uncertain a fresh hole is bored and the line traced on. When experimenting with flow not in the immediate vicinity of the model surface, meshes of fine string or wire coated with sesquichloride of iron are used and pyrogallic acid ejected at known points.

The relative flow indicated in the immediate vicinity of the model is found to extend as regards type quite a distance from the skin, so as regards motion near the hull we need consider only the disturbance close to the bottom, or the lines of flow as they may be called.

Figs. 50 to 59 show lines of flow past the bottom for five pairs of twenty-foot models of five widely varying types of midship section. The proportions, displacements and speeds of the models are given. The large and small models of each type of midship section are similar except as regards ratios of beam and draught to length. These figures are typical and confirmed by investigations of the lines of flow over a number of other models. Perhaps their most notable feature is the remarkably strong tendency of the water to dive under the fore body as it were. In fact, it seems as if the water near the surface forward dives down and crowds away from the hull the water through which the fore part has passed, while aft the water rising up crowds away from the hull the water which was in contact with it near the surface amidships.

4. Kelvin's Wave Patterns and Actual Ship Wave Patterns. — It remains to consider the most striking of the disturbances caused by a moving ship. This is the surface or wave disturbance.

The wave disturbance caused by a ship differs obviously from trochoidal waves, which we have considered.

These latter were considered as an infinite series of parallel crests, each crest line extending to infinity.

We owe to the genius of Lord Kelvin the solution of an ideal problem which applies reasonably well to ship waves. His work in this connection, which may be found in the Transactions of the Royal Society of Edinburgh (Vols. XXV (1904–5) and XXVI (1906)), bristles with difficult mathematics, but his results are comparatively simple.

Suppose we have advancing in a straight line over the surface of a perfect liquid a point of disturbance. What will be the resulting waves? Lord Kelvin's conclusion is that there will be a number of crests, each crest line being represented by

$$(x^2 + y^2)^3 + a^2 (8 y^4 - 20 x^2 y^2 - x^4) + 16 a^4 y^2 = 0$$

where the origin is supposed to travel in the direction of the axis of x with and at the point initiating the disturbance.

The equation above is somewhat simpler in polar coördinates. Transformed it becomes

$$r^4 - a^2 r^2 (1 + 18 \sin^2 \theta - 27 \sin^4 \theta) + 16 a^4 \sin^2 \theta = 0.$$

Fig. 60 shows a single crest line from the above equation. It starts always from o, where it is tangent to the axis of x. It spreads outward and backward to cusps CC, which are on a line making with the axis of x the angle of 19° 28′ whose tangent is $\sqrt{\frac{1}{8}}$ or sine is $\frac{1}{3}$. The tangent at the cusp is inclined 54° 44′ to the axis of x, and the branch CAC of the crest line is perpendicular to the axis of x where it crosses it. The relative heights of various points on the crest as given by Lord Kelvin are indicated in Fig. 60. The fact that the heights at O and CC are infinite shows simply that the formula cannot represent the physical conditions with exactness. It may, however, be an amply close approximation, for by the theory these infinite crest heights extend for but infinitely short distances.

The physical interpretation of the formula is that at OC and C the heights are greatest and the crests the sharpest, so that at these points, if anywhere, breaking water will be found. This conclusion is fully borne out in practice.

The whole wave disturbance due to the initiating point is made up by the super position of a series of crests such as are outlined in Fig. 60, with corresponding intervening hollows. Fig. 61 shows a series of such crest lines. The diverging crest lines cross the transverse crest lines, resulting in an involved surface disturbance. The distance between successive transverse crests along the axis of x is the same as the length of an ordinary trochoidal wave traveling in deep water at the speed of the point of initial disturbance. The heights of successive crests are inversely as the square roots of distances from the origin.

That Lord Kelvin's solution agrees reasonably well with practical results is readily shown by careful scrutiny of the wave disturbances caused by ships and models, which makes it clear that the bow wave system and the stern wave system closely resemble Kelvin wave groups.

The differences are only such as might be expected from the fact that a Kelvin group is an ideal system initiated by forces at a single moving point, while an actual wave group is due to forces spread over the ship's hull.

The heights of the later diverging waves close to the ship appear to be much less in practice than by the Kelvin formula, these crests frequently appearing as mere wrinkles of the surface, and the ship wave patterns vary with proportions of the vessel. Thus narrow deep ships have wave patterns whose transverse features are much more strongly accentuated than those of broad shallow ships.

The wave patterns of ships appear to change somewhat with change of speed and the transverse features appear to be less prominent and important at high speed. According to observations made by Commander Hovgaard, formerly of the Danish Navy, and given by him in a paper before the Institution of Naval Architects at its spring meeting in 1909, the cusp line is usually at an angle less than $19°\ 28'$, most observations of full-sized ships showing it between $16°$ and $19°$, though in one case, that of a Danish torpedo boat, Commander Hovgaard observed a cusp line angle as low as $11°$.

Observations made on models by Commander Hovgaard in the United States Model Basin showed even smaller values of cusp line angles, particularly at relatively high speeds.

But at such speeds the breadth of the basin is not sufficient to allow the cusp line to be determined with accuracy.

For purposes of analysis the most important feature of the Kelvin wave group is the close agreement between its curved transverse crests and a series of transverse trochoidal crests extending from the cusp line on one side to the cusp line on the other.

5. Havelock's Wave Formulæ. — Lord Kelvin's wave formulæ given above are for deep water. Dr. T. H. Havelock has developed formulæ for the wave patterns produced by a traveling disturbance in water of any depth. These will be found in a paper on waves, etc., in the Proceedings of the Royal Society, Vol. 81, 1908.

In a paper on Wave-making Resistance of Ships, Vol. 82, 1909, Dr. Havelock has applied his formulæ to produce practical results. For waves generated by a traveling disturbance in deep water

Havelock's results agree with Kelvin's except that Havelock's formulæ do not require infinite wave heights.

But in shallow water Havelock finds that there is a critical speed, which is, in feet per second, $\sqrt{gh}$, where h is the depth in feet. This is, by the way, the speed of the solitary wave or wave of translation by the trochoidal formulæ.

As the speed increases up to the critical speed the cusp line angle, which was 19° 28′ in deep water, becomes greater and greater until at the critical speed it is 90°. At this speed the wave disturbance reduces to a single transverse wave.

Above the critical speed transverse waves cannot exist. Diverging waves continue however, but instead of being concave the first one is straight at an angle which decreases from 90° with the axis as speed increases beyond the critical speed.

The succeeding diverging waves are convex instead of concave. We shall see later that observed phenomena accompanying the motion of models in shallow water are in accordance with Havelock's theoretical conclusions.

CHAPTER II

Resistance

7. Kinds of Resistance

There are several kinds of resistance and usually all are present in the case of every ship. They will be enumerated here and then taken up separately in detail.

1. **Skin Resistance.** — In the first place, water is not frictionless. Its motion past the surface of the ship involves a certain amount of frictional drag, the resistance of the surface involving an equal and opposite pull upon the water.

This kind of resistance is conveniently denoted by the term Skin Resistance. It is nearly always the most important factor of the total resistance.

2. **Eddy Resistance.** — While Skin Resistance is accompanied by eddies or whirls in the water near the ship's surface, the expression Eddy Resistance is used for a different kind of resistance. The motion through the water of a blunt or square stern post or of a short and thick strut arm, etc., is accompanied by much resistance and the tailing aft of a mass of eddying confused water. Such resistance is designated Eddy Resistance. With proper design it is in most cases but a minor factor of the total resistance.

3. **Wave Resistance.** — A far more important factor, which though usually second to the Skin Resistance is in some cases the largest single factor in the total resistance, is the resistance due to the waves created by the motion of the ship. It is called for brevity the Wave Resistance.

We have seen that the motion of a ship through the water is accompanied by the production of surface waves. These absorb energy in their production and propagation, and this energy is communicated to them from the ship, being derived from the Wave Resistance.

4. Air Resistance. — Finally, we have the Air Resistance, which is, as its name implies, the resistance which the air offers to the motion of the ship through it. The Air Resistance is seldom large. It is, however, by no means always negligible.

5. Comparative Importance of Skin and Wave Resistance. — Considering the two main factors of resistance, namely, Skin Resistance and Wave Resistance, experience shows that for large vessels of very low speed the Skin Resistance may approach 90 per cent of the total. For ordinary vessels of moderate speed, it is usually between 70 and 80 per cent of the total. As speed increases, the Wave Resistance becomes a more and more important factor, until, in some cases of vessels pushed to speeds very high for their lengths, the Skin Resistance may be only some 40 per cent of the total, the Wave Resistance being in the neighborhood of 60 per cent. For such vessels as high-speed steam launches the Wave Resistance may be even more than 60 per cent of the total, but for vessels of any size it is seldom advisable to adopt a design where the Wave Resistance is as great as 50 per cent of the total.

Features which tend to decrease Wave Resistance tend to increase Skin Resistance, and here, as in so many other matters, the naval architect must adopt a compromise dictated by the special considerations affecting the particular case.

8. Skin Resistance

1. William Froude's Experiments. — The determination of the Skin Resistance of ships is based entirely upon the experimental determination of the frictional resistance of thin comparatively small planes moving endwise through the water. The classical experiments in this connection were made by Mr. William Froude many years ago and are recorded in the Proceedings for 1874 of the British Association for the Advancement of Science.

Mr. Froude used boards $\frac{3}{16} \times 19$ inches, of various lengths up to 50 feet and coated with various substances, which were towed at various speeds not exceeding eight knots in a tank of fresh water 300 feet long, their resistance being carefully measured. Mr. Froude summarized his experimental results in the following table:

RESULTS OF WILLIAM FROUDE'S EXPERIMENTS UPON SKIN FRICTION

For Speed of 600 Feet per Minute

Nature of Surface.	Length of Surface or Distance from Cutwater.											
	2 feet.			8 feet.			20 feet.			50 feet.		
	A	B	C	A	B	C	A	B	C	A	B	C
Varnish.........	2.00	.41	.390	1.85	.325	.264	1.85	.278	.240	1.83	.250	.226
Paraffin.........	1.95	.38	.370	1.94	.314	.260	1.93	.271	.237	....	....	....
Tinfoil..........	2.16	.30	.295	1.99	.278	.263	1.90	.262	.244	1.83	.246	.232
Calico..........	1.93	.87	.725	1.92	.626	.504	1.80	.531	.447	1.87	.474	.423
Fine sand......	2.00	.81	.690	2.00	.583	.450	2.00	.480	.384	2.06	.405	.337
Medium sand...	2.00	.90	.730	2.00	.625	.488	2.00	.534	.465	2.00	.488	.456
Coarse sand....	2.00	1.10	.880	2.00	.714	.520	2.00	.588	.490	....	....	....

In the above, for each length stated in the heading —

Column A gives the power of the speed according to which the resistance varies.

Column B gives the mean resistance in pounds per square foot of the whole surface for a speed of 600 feet per minute.

Column C gives the resistance in pounds, at the same speed, of a square foot at the distance abaft the cutwater stated in the heading.

It appears, then, that the Frictional Resistance of a plane surface can be represented by the formula $R_f = fSV^n$, where S is the total surface of the plane, f is its coefficient of friction, V is its speed, and n an index giving the power of V according to which R_f is increasing. The table below repeats the values of n in Froude's table above and gives the values of f from columns B and C when we express speed in knots, S being expressed in square feet.

RESULTS OF WILLIAM FROUDE'S EXPERIMENTS UPON SKIN FRICTION

Reduced for Speeds in Knots

Nature of Surface.	\multicolumn{3}{c}{2 feet.}			8 feet.			20 feet.			50 feet.		
		f			f			f			f	
	n	B	C	n	B	C	n	B	C	n	B	C
Varnish......	2.00	.0117	.0111	1.85	.0121	.0098	1.85	.0104	.0089	1.83	.0097	.0087
Paraffin......	1.95	.0119	.0115	1.94	.0100	.0083	1.93	.0088	.0077	....	....	....
Tinfoil.......	2.16	.0064	.0063	1.99	.0081	.0076	1.90	.0089	.0083	1.83	.0095	.0090
Calico........	1.93	.0281	.0234	1.92	.0106	.0166	1.89	.0184	.0155	1.87	.0170	.0152
Fine sand.....	2.00	.0231	.0197	2.00	.0166	.0128	2.00	.0137	.0110	2.06	.0104	.0086
Medium sand.	2.00	.0257	.0208	2.00	.0178	.0139	2.00	.0152	.0133	2.00	.0139	.0130
Coarse sand	2.00	.0314	.0251	2.00	.0204	.0148	2.00	.0168	.0140	....	....	....

2. Variation in Coefficient and Index of Friction.—The coefficient of friction f is affected by a number of circumstances. The tables preceding show a variation with nature and length of surface. It also varies slightly with temperature, falling off as temperature increases, and it varies, of course, with the nature of the fluid. For the small variations of density from fresh to salt water f is taken to vary directly as the density.

As to the index n, it is seen that for rough surfaces it remains at the value 2.00, while for smooth hard surfaces it falls off with increase of length, reaching the value 1.83 for planes 50 feet long. The diminution in f and that in n as length is increased are both due to the same cause, namely, the fact that the rear portion of a plane moves through water which has a forward motion caused by the friction of the front portion of the plane. Froude's conclusion that for a plane with smooth hard surface the index n has a value of 1.83 or thereabouts has been fully confirmed by other experiments. Prior to Froude it was always considered that the frictional resistance of a plane surface would vary as the square of the speed.

This seems natural, and most experiments on the loss of head of water flowing through pipes show that the resistance to flow varies as the square of the speed. The conditions are, however, very different. In the case of the pipe we consider the average velocity of flow over the cross section of the pipe, which is necessarily the same from end to end, and its ratio to the rubbing velocity of the water close to the walls of the pipe is practically constant. In the case of the plane, the rubbing velocity steadily falls off along the plane.

While the frictional index 1.83 for long smooth surfaces does not differ greatly from 2, the corresponding curve is far below the parabola corresponding to the index 2. Thus the ratio $V^{1.83} \div V^2$, which is unity for $V = 1$, is .761 for $V = 5$, is .676 for $V = 10$ and .609 for $V = 20$. This ratio falls off more and more slowly as speed is increased. Thus, in passing from $V = 1$ to $V = 20$ it falls off from 1.000 to .609, while to reduce it to .500 the speed must increase to $V = 59$.

3. **Frictional Resistance of Ships Deduced from Plane Results.** — In order to apply the results for friction of planes to the frictional resistance of ships, it is necessary first to extend the experimental results for short planes to long surfaces, the lengths of actual ships. This has been done by Froude and Tideman, by extending the curves of index, coefficient, etc., for the short planes experimented with. While this extension is speculative to some extent, it does not appear that it is likely to be seriously in error.

Then it is assumed that the frictional resistance of the wetted surface of a ship is the same as the frictional resistance of a plane of the same length and total surface moving endwise through the water with the speed of the ship. This assumption is necessarily an approximation. The water level changes around a ship under way, changing the area of wetted surface; and, owing to stream line action, the velocity of flow over the surface is at some places less, at others greater, than it would be over the plane surface. The assumption made, however, is practically necessary, and is a reasonably close approximation to actual facts.

Finally, it is necessary to assume that the frictional quality of the ship's surface is the same as that of our experimental planes.

From experiments made in the Italian model basin and elsewhere it may be concluded that the frictional resistance of a smooth hard surface is not materially affected by the variety of paint with which it is covered. But Froude's experiments show that friction is powerfully affected by roughness of surface. For a 50-foot plane covered with calico or medium sand and towed at 600 feet per minute, or about 6 knots, Froude found a frictional resistance nearly double that of a varnished plane of the same size. The calico surface had an index but little greater than the varnished surface, so its friction would remain in nearly constant ratio to that of the varnished surface. The medium sand, however, had a greater index. This results in a much greater relative increase at high speeds. Thus, using Froude's coefficients, the ratio between medium sand and varnish, which is 1.43 at one knot, becomes 2.12 at 10 knots, 2.38 at 20 knots, and 2.56 at 30 knots.

The relatively enormous increase of frictional resistance with fouling is well known, but we have very little quantitative information as to the difference as regards frictional quality even between the smoothest possible steel ship and one whose bottom, while acceptably fair, is not ideally smooth.

It would be very desirable to narrow the gaps which we must now bridge by assumptions in connection with frictional resistance from the results of experiments on large and long planes of various surfaces made in open water at high speeds. Such experiments would, however, be very difficult. It would be very hard to tow such planes straight.

Pending such experiments, we must rely upon coefficients deduced from the small scale experiments.

4. R. E. Froude's Frictional Constants. — Mr. R. E. Froude, in a paper in 1888, before the Institution of Naval Architects, has supplemented the British Association paper of his father, Mr. William Froude, by data of coefficients and constants used by him, from which Table V of Froude's Frictional Constants has been computed.

It will be noted that as regards paraffin surfaces the table differs slightly from Mr. William Froude's results, obtained in 1872.

Mr. R. E. Froude states that as regards the paraffin in use in 1888 it appeared identical in frictional quality with a smooth painted or varnished surface.

5. **Tideman's Frictional Constants.** — Closely following the elder Froude's classical experiments of 1872, Herr B. Tideman, Chief Constructor of the Dutch Navy, made a number of similar experiments, from which he deduced a complete set of frictional constants. These are given in Table VI. The most important are those for "Iron Bottom — Clean and Well Painted." These are comparable with Froude's constants, and it will be noted that they are slightly greater.

For varnished planes 20 feet long, Froude's constants agree very closely with results of careful experiments at the United States Model Basin; but for full-sized ships it is considered preferable to use Tideman's coefficients, simply because they are slightly larger, and hence make some allowance for the imperfections of workmanship found in practice. At the United States Model Basin, it is the practice, when dealing with vessels more than 100 feet long, to use the Tideman values of f, but the index 1.83 instead of 1.829, as given by Tideman. This increases Tideman's results by negligible amounts.

6. **Law of Comparison not Applicable to Frictional Resistance.** — Having concluded, then, that we should represent the frictional resistance of a ship by $R_f = fSV^{1.83}$, where R_f is frictional resistance in pounds, f is a coefficient varying slightly with length, S is wetted surface in square feet and V is speed in knots, let us see whether we can apply the Law of Comparison to resistance following the formula.

Let R_{1f}, f_1, S_1, V_1 refer to one ship, R_{2f}, f_2, S_2, V_2 to a similar ship.

Then $R_{1f} = f_1 S_1 V_1^{1.83}$. $R_{2f} = f_2 S_2 V_2^{1.83}$. Let the ratio of linear dimensions of the two ships be λ and let V_2 and V_1 be in the ratio $\sqrt{\lambda}$, as required by the Law of Comparison.

Then
$$\frac{R_{2f}}{R_{1f}} = \frac{f_2}{f_1} \frac{S_2}{S_1} \left(\frac{V_2}{V_1}\right)^{1.83}.$$

Now
$$\frac{S_2}{S_1} = \lambda^2$$

and we have made $\dfrac{V_2}{V_1} = \sqrt{\lambda} = \lambda^{\frac{1}{2}}$.

Then at corresponding speeds

$$\dfrac{R_{2f}}{R_{1f}} = \dfrac{f_2}{f_1}\lambda^2 \quad \lambda^{\frac{1.83}{2}} = \dfrac{f_2}{f_1} \quad \lambda^{2.915}.$$

But to satisfy the Law of Comparison we should have at corresponding speeds $\dfrac{R_{2f}}{R_{1f}} = \lambda^3$. We see, then, that frictional resistance does not follow the Law of Comparison, and hence we cannot deduce the frictional resistance of a full-sized ship from that of a model. Thus suppose we had a vessel 500 feet long of 12,500 tons displacement and 39,000 square feet wetted surface. A similar 20-foot model would have 62.4 square feet of wetted surface. If the speed of the ship were 20 knots, the corresponding model speed would be 4 knots = $20\sqrt{\dfrac{20}{500}}$.

Using Froude's coefficient and 1.83 index, the frictional resistance of the 20-foot model would be $.01055 \times 62.4 \times 4^{1.83} = 8.3218$ pounds. If the Law of Comparison held, this would make R_f for the full-sized ship at 20 knots $8.3218\,(25)^3 = 130{,}028$ pounds. But using Froude's coefficient of friction we have for the full-sized ship $R_f = 39{,}000 \times .00880 \times 20^{1.83} = 82{,}495$ pounds, and using Tideman's coefficient $R_f = 84{,}745$ pounds.

It is seen, then, that the Skin Friction, as we calculate it, falls far short in practice of what it would be if the Law of Comparison were applicable to it.

7. **Air Disengaged around Moving Ships.** — There is one phenomenon generally accompanying the motion of a full-sized ship which seldom manifests itself in model experiments. As a fast ship moves through the water, it is seen that the water in the immediate vicinity of the skin plating, particularly aft of the center of length, has a great many air bubbles. The air is either disengaged from water in which it is entrained by the reduction of pressure in frictional eddies, or it is carried down and along the ship as a result of breaking water toward the bow. However

produced, its presence must reduce the density of a layer of water covering a large portion, if not all, of the surface of the bottom, and it would seem, at first, that there should be a corresponding reduction of friction. It is in fact a favorite dream of inventors to deliver air around the outside of a ship so that the immersed surface will be surrounded by a film of air instead of water. Could this result be accomplished, it would undoubtedly result in a great reduction of skin friction. But air released under water persists in forming into globules, not films. Experiments have been made at the United States Model Basin by pumping air around a model through a number of holes near the bow and out through narrow vertical slots in the forward portion of a 20-foot friction plane. The results of these experiments were that for the model the resistance was always materially increased when the air was pumped out. In this case the air came out through holes and promptly formed globules. In the case of the friction plane the air came out in a thin film which spread aft. At speeds of 12 to 16 knots, when the films of air on each side visibly extended over perhaps a third of the plane, the resistance was almost exactly the same as when no air was pumped. At speeds below 12 knots the resistance was greater when the air was pumped.

It is possible that for vessels of the skimming-dish or other abnormal type the efforts of inventors to reduce resistance by means of air cushions may be successful, but there is little doubt that no matter how much air may be forced into the water around a ship of ordinary type, practically none of it remains in contact with the ship's surface. That is covered always by a film of solid water. The air forms globular masses or bubbles and never touches the surface of the hull. While in an actual ship the air bubbles naturally appearing must somewhat reduce the density of some of the liquid around the bottom, it appears likely that, to reduce skin friction materially, this reduction of density would have to extend to a much greater distance from the hull than is usually the case and that in practice the evolution of air found probably increases the resistance by an uncertain amount. This uncertainty could be removed by frictional experiments upon planes of such size and nature of surface as to be closely comparable to actual ships.

8. Effect of Foulness upon Skin Resistance.

— In design work we usually deal primarily with clean bottoms. When vessels become foul by the accumulation of marine growths such as grass and shellfish the Skin Resistance is much increased. Fig. 62 illustrates the effect of change of surface upon Skin Resistance.

Froude's experimental results for five surfaces are extended by his formula to high speeds. The two smooth hard surfaces — varnish and tinfoil — are nearly the same. But a surface covered with calico shows about double as much resistance, and surfaces covered with fine or medium sand show more than double the resistance of the varnished surface at speeds above 20 knots. When we reflect that in the most extreme cases of fouling a vessel's bottom may have a complete incrustation of shellfish it is easy to realize that fouling may result in Skin Resistance four or five times that of the clean ship.

Of course in practice such fouling is permitted only under exceptional circumstances, vessels in service being docked at intervals. But even in cool waters where fouling usually goes on rather slowly a vessel three or four months out of dock is liable to have an increase of 20 per cent or more in Skin Resistance, and in tropical waters the increase of resistance is greater.

Foulness is usually gauged by the loss of speed, which tends to mask the great increase of Skin Resistance. Thus a loss of two knots of speed for the same power means in the case of a vessel originally of moderate speed an increase of about 100 per cent in Skin Resistance.

When in design work it is necessary to allow for the effect of fouling it is usually done indirectly by providing a margin of speed with a clean bottom equal to the loss to be expected from fouling. This loss must be estimated from previous experience with vessels in the service under consideration.

9. Eddy Resistance

As already stated, Eddy Resistance is a minor factor in the case of most ships and cannot be determined separately by experiment. It is possible, however, to get a reasonably good idea of the laws of Eddy Resistance by experiments with planes, sections of strut arms, and similar appendages.

1. **Flow Past a Thin Plane Producing Eddy Resistance.** — Fig. 63 shows a section through a plane AB and a stream of water flowing past it, and indicates, diagrammatically, what happens. The plane is inclined at an angle α to the direction of undisturbed flow; K is the dividing point of the stream. On one side of K the water flows around the corner at A. On the other side it flows by B. The position of K depends upon the angle α. In front of the plane there is practically perfect stream motion, as indicated. The velocity of the water is checked, with corresponding increase of pressure, but there is no discontinuity. In the rear of the plane, however, the conditions are different. The water breaks away at A and B, and there is found behind the plane a mass of confused eddying water, whose pressure must be reduced below the normal pressure due to depth below the surface, but in a more or less erratic manner.

2. **Rayleigh's Formulæ for Eddy Resistance.** — The total Eddy Resistance of the plane would then be due to a front pressure and a rear suction. These are evidently but little dependent upon each other. The front pressure has been investigated theoretically by assuming a smooth solid inserted behind the plane, so that the water has perfect stream motion throughout. The resulting formulæ as deduced by Lord Rayleigh are as follows:

$$P_n' = \frac{2\pi \sin \alpha}{4 + \pi \sin \alpha} \frac{w}{2g} Av^2,$$

$$\frac{AK}{AB} = \frac{2 + 4 \cos \alpha - 2 \cos^3 \alpha + (\pi - \alpha) \sin \alpha}{4 + \pi \sin \alpha}.$$

In these formulæ P_n' is normal pressure or total pressure perpendicular to the front face of the plane, α is the angle the plane makes with the direction of motion, w is the weight per cubic foot of the water, g is the acceleration due to gravity, A is area of plane in square feet and v is its velocity in feet per second.

It may be noted that at K, where the water is brought completely to rest, the excess pressure is $\frac{w}{2g} v^2$. If this pressure were over the whole plane, the total normal front pressure would be $\dfrac{w}{2g} A v^2$.

The fraction $\dfrac{2\pi \sin \alpha}{4 + \pi \sin \alpha}$ is, then, the ratio between the front pressure and the pressure due to velocity multiplied by the area of the plane. This fraction is, as might be expected, a maximum for $\alpha = 90°$. Its value, then, is $\dfrac{2\pi}{4+\pi}$ or .88. This is materially less than unity, and as α decreases the fraction soon begins to fall off rapidly. Fig. 64 shows curves of the ratio $\dfrac{2\pi \sin \alpha}{4 + \pi \sin \alpha}$ and the ratio $\dfrac{AK}{AB}$ plotted on α.

The front pressure by Rayleigh's formula follows the Law of Comparison. For suppose we have two similar planes at the same angle. If P_1 denote the front pressure on No. 1 and P_2 the front pressure on No. 2,

$$P_1 = \dfrac{2\pi \sin \alpha}{4 + \pi \sin \alpha} \dfrac{w}{2g} A_1 v_1^2, \qquad P_2 = \dfrac{2\pi \sin \alpha}{4 + \pi \sin \alpha} \dfrac{w}{2g} A_2 v_2^2.$$

Whence $\dfrac{P_1}{P_2} = \dfrac{A_1 v_1^2}{A_2 v_2^2}$. Now if λ denote ratio of linear dimensions, $A_1 = \lambda^2 A_2$ and for corresponding speeds $v_1^2 = \lambda v_2^2$. Then at corresponding speeds $\dfrac{P_1}{P_2} = \lambda^3$, or Froude's Law is satisfied.

For salt water $\dfrac{w}{2g} = 1$ practically. Furthermore it is desirable to reduce all speeds to knots, denoted by V. When this is done Rayleigh's formula for front face pressure may be written

$$P_n' = \dfrac{5.705 \sin \alpha}{1.273 + \sin \alpha} A V^2.$$

3. Joessel's Experiments and Formulæ for Eddy Resistance. — When we come to consider the total normal resistance of an inclined plane moving through water we are compelled to rely upon semi-empirical formulæ derived by experiments.

It is impossible to reduce the resistance due to confused eddying behind the plane to mathematical law. The ground has never been adequately covered experimentally, and it is as a matter of fact a question whose accurate experimental investigation presents many difficulties.

M. Joessel made experiments with small planes 12 inches by 16 inches in the river Loire at Indret, near Nantes, about 1873. The maximum current velocity was only about $2\frac{1}{2}$ knots. Joessel's results may be expressed as follows:

If l denote the breadth of a plane in the direction of motion making the angle α with the direction of flow and x the distance of the center of pressure from the leading edge,

$$x = (.195 + .305 \sin \alpha) \, l.$$

If P_n denote total normal force due to pressure in front and suction in rear, we have for area A in square feet and velocity V in knots

$$P_n = \frac{7.584 \sin \alpha}{.639 + \sin \alpha} A V^2.$$

4. John's Analysis of Beaufoy's Eddy Resistance Experiments. — Mr. A. W. John in an interesting paper on "Normal Pressures on Thin Moving Plates," before the Institution of Naval Architects in 1904, has analyzed Colonel Beaufoy's experiments of 1795 with square plates of about three square feet area (double plates abreast one another about 8 feet apart and 3 feet below the surface) and shown that the results present the following peculiar features. Up to about 30 degrees inclination the normal pressure increases linearly, and from 30 degrees to 90 degrees it remains almost constant. The same result has been found by various recent experiments with planes in air. It appears to be characteristic of squares, circles and rectangles approaching the square, and is not so pronounced in the case of long narrow rectangles moving perpendicular to the long side.

Beaufoy's results as plotted by John may be approximately expressed by a semi-empirical formula of the same form as Rayleigh's formula,

$$P_n = \frac{A \sin \alpha}{B + \sin \alpha} A V^2.$$

This may be made to coincide at two points with the experimental results. We have

For coincidence at $\alpha = 90°$ and $\alpha = 10°$ $A = 5.20$, $B = .557$
For coincidence at $\alpha = 90°$ and $\alpha = 15°$ $A = 4.63$, $B = .389$
For coincidence at $\alpha = 90°$ and $\alpha = 20°$ $A = 4.08$, $B = .223$

It is reasonable to take the values for $\alpha = 15°$. We then have formula derived from Beaufoy,

$$P_n = \frac{4.63 \sin \alpha}{.389 + \sin \alpha} A V^2.$$

5. Stanton's Eddy Resistance Experiment. — Dr. T. E. Stanton has recently made experiments with very small plates of 2 square inches area in an artificial current of water of 4 knots velocity. His results are published and discussed in a paper of April 2, 1909, before the Institution of Naval Architects. He found the same phenomenon developed by John's analysis of Beaufoy's experiments, namely that the normal pressure on a square plate rises almost linearly to an angle of 35° or so and then does not change much from 35° to 90°. For a plate whose length in the direction of motion was twice its width there was a pronounced hump at about 45°, the normal pressure at this inclination being 13 or 14% greater than at 90°. For a plate of length in the direction of motion but one-half its width the hump feature was not so pronounced and was strongest at an inclination below 30°.

6. Formulæ for Eddy Resistance of Normal Plates Compared. — When $\alpha = 90°$, or the plane moves normally to itself, we have

By Rayleigh's formula: — Pressure on front face $= P_n' = 2.51\, A V^2$
By Joessel's formula: — Total normal force $\quad = P_n = 4.63\, A V^2$
By formula from Beaufoy's results, $\qquad P_n = 3.33\, A V^2$
From Stanton's results, $\qquad P_n = 3.42\, A V^2$

It is probable that Rayleigh's formula expresses quite closely the resistance of a square stem for instance. If we adopt Joessel's formula, which gives the largest resistance, and deduct the front face pressure, we would have for rear suction $P_r = 2.12\, A V^2$. This formula will probably give an outside value for resistance such as that of a square stern post.

7. Formulæ for Eddy Resistance of Inclined Plates Compared.

— For small values of α it is convenient to use a formula of the form $P_n = C \sin \alpha\, AV^2$. If we choose C to correspond to P_n from the complete formula for an angle of 15 degrees we can simplify Rayleigh's formula, etc., for use up to angles of 30° or so. Stanton's results are already expressed in this simple form, and William Froude has a formula of this type expressing normal force for small angles of inclination.

Rayleigh's formula becomes	$P_n' = 3.73 \sin \alpha\, AV^2$
Joessel's formula becomes	$P_n = 8.45 \sin \alpha\, AV^2$
Formula from Beaufoy becomes	$P_n = 7.15 \sin \alpha\, AV^2$
Froude's formula becomes	$P_n = 4.85 \sin \alpha\, AV^2$
Stanton's formula for a square plate becomes	$P_n = 5.13 \sin \alpha\, AV^2$
Stanton's formula for a plate twice as broad as long becomes	$P_n = 7.70 \sin \alpha\, AV^2$

The above formulæ are not very consistent with each other. The question of planes advancing at various angles through water is in need of a complete and accurate experimental investigation. It may be noted that Stanton's plane twice as broad as long approaches somewhat the proportions of an ordinary rudder of barn-door type, and his coefficient for such a plate agrees well with Joessel's results, which have been used a good deal for rudder work in France. In England, the so-called Beaufoy's formula has been much used for rudders. This gives $P_n = 3.2 \sin \alpha\, AV^2$, a value much below that from Joessel's formula. But in using this formula, the center of pressure is assumed to be at the center of figure instead of forward of it as by Joessel's formula for center of pressure. The net result is that the English formula gives a twisting moment on the rudder stock at usual helm angles only about 30 per cent less than that derived from Joessel's complete formulæ. This is for ordinary rudders. For partially balanced rudders the difference is somewhat less.

Experiments with rudders have indicated normal pressures on them materially less than and sometimes but a fraction of what would be given by Joessel's formula when V was taken as the speed of the ship. But the true speed of a rudder through the water in its vicinity is nearly always less and often much less than the speed of the ship, and there are other conditions wherein a rudder differs very much from a detached plate.

8. Eddy Resistance Formulæ Applicable to Ships. — All things considered, it seems well, pending more complete experimental investigation, to use for a plane Rayleigh's formula for front face resistance and Joessel's for total resistance.

Then we would have for a square stem, the end of a bow torpedo tube, and similar fittings having head resistance only,

$$P_n' = 2.5 \, AV^2.$$

For square stern posts and similar objects $P_r = 2.1 \, AV^2$, and for scoops, square or nearly square to the surface of the ship, and similar fittings, $P_n = 4.6 \, AV^2$. In these formulæ A is area in square feet, V is speed of the ship in knots and P_n, etc., are in pounds.

It is probable that these formulæ would nearly always overestimate the resistance concerned, but as the resistances to which they apply constitute a very small portion of the total in most cases, it is not necessary to estimate them with great accuracy and it is advisable to overestimate rather than underestimate them.

The resistance of struts is largely eddy resistance, but methods for dealing with them will be considered in connection with appendages.

9. Formula for Eddy Resistance behind Plate has Limitations. — In connection with the formula suggested for rear suction, namely $P_r = 2.1 \, AV^2$, it should be pointed out that this cannot apply as speed is increased indefinitely.

Consider a plane of one square foot area immersed 10 feet say. The pressure on its rear face, allowing 34 feet of water as the equivalent of the atmospheric pressure and taking water as sea water weighing 64 pounds per square foot, would be $44 \times 64 = 2816$ pounds. Evidently there is maximum rear suction when there is a vacuum behind and no pressure on the rear face. Hence 2816 pounds is the maximum possible rear suction. By the formula, if $P_r = 2816 = 2.1\ V^2$, $V^2 = \dfrac{2816}{2.1} = 1341$, $V = 36.62$. Then the formula obviously cannot apply beyond $V = 36.62$. Even if the constant 2.1 is too great we will still in time reach a speed where any formula of this type will give a rear suction equal to the original forward pressure. Any formula which assumes that suction increases indefinitely as the square of the speed must then be regarded as expressing not a scientific fact but a convenient semi-empirical approximation to the actual facts over the range of speeds found in practice.

10. Wave Resistance

In discussing the disturbances of the water by a ship we have given some consideration to the waves produced. To maintain these waves, energy must be expended which can come only from the ship. That portion of the ship's resistance which is absorbed in raising and maintaining trains of waves is conveniently called Wave Resistance.

1. **Bow and Stern System.** — The tendency is toward the formation of two distinct series of waves — one initiated at the bow and conveniently called the Bow Wave System and the other initiated at the stern and called the Stern Wave System. The Stern Wave System, however, makes its appearance in water already more or less disturbed by the Bow Wave System and hence the ultimate wave disturbance is compounded of the two systems.

When considering Kelvin's wave system as illustrated diagrammatically in Figs. 60 and 61, we saw that it was made up of transverse crests and diverging crests, the transverse crests being but little curved and extending to the cusp line on each side. For a given speed the length between successive transverse crests is the same as the trochoidal wave length for the same speed.

It is evidently a reasonable approximation under the circumstances to substitute for the actual wave systems ideal systems composed of traverse trochoidal waves extending out to the cusp lines of Kelvin's waves and each wave of uniform height such that energy of the ideal systems is the same as that of the actual systems.

Consider first the bow system. To maintain this system there must be communicated to it while the ship advances the length of one wave energy proportional to the energy of one wave length.

If we denote by l the length from crest to crest of the trochoidal wave, by b its mean breath and by H its height, w being the weight of water per cubic foot, we know from the trochoidal wave formulæ that the energy per wave length is proportional to $wblH^2$. Now the external energy communicated to the system by the wave resistance R_w while the ship traverses a wave length l is proportional to $R_w l$. Hence $R_w l$ is proportional to $wblH^2$ or $R_w \propto wbH^2$. A similar formula applies to the stern wave resistance.

2. **Resultant Wave System.** — The actual wave resistance is due to the wave system formed by compounding the bow and stern wave systems. To determine the resultant system we compound the bow and stern wave systems by the formulæ for compounding trochoidal waves.

In order to determine the resultant of the two separate wave systems of the same length advancing in the same direction, we need to know the distance between crests, and it is advisable to consider the first crest of each system. The first crest of the bow wave system will be somewhat abaft the bow and the first crest of the stern wave system somewhat abaft the stern. Their positions and the distance between them will vary with speed. Call the distance between them the wave-making length of the ship and denote it by mL, where m is a coefficient varying slightly with speed and, as we shall see, somewhat greater than unity. Now, if V is the speed of the ship in knots, the bow wave length l in feet is $.5573\ V^2$. The distance between the first stern system crest and the bow system crest next ahead of it is evidently the remainder after subtracting from mL the lengths of the complete waves, if any, between the first bow crest and the first stern crest. Let there be n such waves and let the distance between the first stern system crest and the bow system crest next forward of it be ql, where l is the wave length. Then $mL = (n + q)\, l = (n + q)\, .5573\ V^2$, where n is a whole number and q is a fraction. In the compound wave formula we need to know $\cos \dfrac{a}{R}$ or $\cos \dfrac{2\pi a}{l}$. Now, a in the above is evidently the same as ql. Then $\cos \dfrac{2\pi a}{l} = \cos \dfrac{2\pi ql}{l} = \cos 2\pi q$. Now n being a whole number, $\cos 2\pi q = \cos 2\pi (q+n)$ and $q+n = \dfrac{mL}{.5573\,V^2}$. Hence $\cos \dfrac{2\pi a}{l} = \cos \dfrac{2\pi mL}{.5573\,V^2} = \cos \dfrac{360°}{.5573} \dfrac{m}{\frac{V^2}{L}}$.

Denote $\dfrac{V^2}{L}$ by c^2. Then $\cos \dfrac{2\pi a}{l} = \cos \dfrac{m}{c^2} 646°$.

The whole bow system is not superposed upon the stern system, but only the inner portion, since the natural bow system extends transversely to a greater distance than the natural stern system.

Let H_1 denote the height of the natural bow system when it has spread to a given breath b, H_2 the height of the natural stern system when it has spread to the same breath. Let kH_1 denote the height of the natural bow system where the stern system has spread to the breath b. Suppose its breath then is b'. Since it has lost no energy $bH_1^2 = b'k^2H_1^2$.

Then the energy per wave length of the compound system resulting from the superposition of a portion of the bow system of breath b upon the whole stern system of breath b is measured by

$$lb\left[k^2H_1^2 + H_2^2 + 2\,kH_1H_2 \cos \dfrac{m}{c^2} 646°\right].$$

The energy of the portion of the bow system beyond the stern system and not compounded is measured by

$$l\,(b' - b)\,k^2H_1^2 = l\,(b'k^2H_1^2 - bk^2H_1^2) = lb\,(H_1^2 - k^2H_1^2),$$

since $b'k^2H_1^2 = bH_1^2$. Adding the above expressions for partial energies the total energy per wave length is measured by

$$lb\left(H_1^2 + H_2^2 + 2\,kH_1H_2 \cos \dfrac{m}{c^2} 646°\right).$$

Whence the wave-making resistance is proportional to

$$b\left(H_1^2 + H_2^2 + 2\,kH_1H_2 \cos \dfrac{m}{c^2} 646°\right).$$

Now, b being an arbitrary convenient constant width, we can say that the wave-making resistance R_w is proportional to

$$H_1^2 + H_2^2 + 2\,kH_1H_2 \cos\frac{m}{c^2}646°.$$

3. General Formula Connecting Wave Resistance and Speed. — The above expression for wave resistance is of little quantitative value without knowledge of coefficients appropriate to all cases, and for practical use in estimating wave resistance there are methods more desirable than the use of a formula, but the expression is of value in enabling us to realize the general nature of the variation of wave resistance with speed.

As a step in this direction we need to know the connection between H_1 and H_2 and the speed.

We know that in perfect stream motion the excess of pressure near the bow is proportional to the square of the speed. If, then, the wave height were proportional to the excess pressure, which it must be approximately, since the surface pressure does not change, we would have H_1 proportional to V^2. Similarly H_2 would be proportional to V^2, and we would have as the general expression for R_w the wave resistance,

$$R_w = V^4\left(A^2 + B^2 + 2\,kAB \cos\frac{m}{c^2}646°\right).$$

The coefficients A and B are not constant. There are two main sources of variation. If the bow wave height were always proportional to the excess bow pressure as speed increases, A would not vary on this account. It seems probable that at moderate speeds when wave resistance first becomes of importance the bow wave height does vary as the excess pressure, but as speed increases a greater proportion of the stream line pressure is absorbed in accelerating the water aft in stream line flow and a less proportion in raising the water level. The same reasoning applies to the stern wave, so, from this point of view, we would expect A and B to be approximately constant at low and moderate speeds and to fall off steadily at high speeds.

There is another important source of variation in A and B. Suppose we have a vessel 400 feet long. Then the length of the fore body is 200 feet. At 13½ knots the length of the bow wave from crest to crest is very nearly 100 feet; at 27 knots it is 400 feet. Then at 13½ knots the bow wave is formed by the forward quarter of the ship, as it were, while at 27 knots the whole forward half of the ship must come into play. The result is, of course, a modification of A and B with speed. There appears to be a critical speed at which the wave length and the wave motion and pressures are in step, as it were, with the ship, and the wave is exaggerated. This may be called the speed of wave synchronism. Broadly speaking, we may say that for fine models of cylindrical coefficient below .55 the speed of wave synchronism in knots is above $\sqrt{L}$, while for full-ended models of cylindrical coefficient above .6 the speed of wave synchronism is below $\sqrt{L}$. We may expect to find a rapid rise of A and B as we approach the speed of wave synchronism and a less rapid falling off as we pass beyond it.

Consider now the coefficient k in the formula

$$R_w = V^4 \left(A^2 + B^2 + 2\,kAB \cos \frac{m}{c^2} 646° \right)$$

At low speeds k is evidently zero, since observation shows that at low speeds the bow disturbance has spread out abreast the stern to a distance where it is not affected one way or the other by the stern disturbance. As the speed increases, however, more and more of the bow wave energy is found in the vicinity of the stern and k may be expected to become greater and greater. It is also a matter of observation that for narrow deep models the transverse features of the bow wave are accentuated, and hence for such models k will, other things being equal, be greater than for broad shallow models, since it is the transverse portion of the bow system which is available for combination with the stern system.

Consider, now, finally, the term $\cos \frac{m}{c^2} 646°$. This expression is equal to $+1$ when $\frac{m}{c^2} 646° = 360°$ or any multiple of $360°$. It is equal to -1 when $\frac{m}{c^2} 646° = 180°$ or $180° +$ any multiple of $360°$. The quantity m is approximately constant for a given ship, though it increases somewhat with the speed. It also appears to increase somewhat with fullness from ship to ship. A fair average value of m would seem to be about 1.15 for speeds where humps and hollows are of importance. For lower speeds m approaches 1. Fig. 65 shows for $m = 1.15$ a curve of $\cos \frac{m}{c^2} 646°$ plotted upon c or $\frac{V}{\sqrt{L}}$. It is seen that at low speeds maxima and minima succeed one another very rapidly. Each maximum corresponds to a "hump" in the curve of residuary resistance and each minimum to a "hollow."

Humps and hollows on actual resistance curves do not manifest themselves, however, in accordance with Fig. 65. The varying term is $kAB \cos \frac{m}{c^2} 646°$, and since in most cases at low speeds k is so small as to be practically negligible, we find in practice that the first important hump usually appears for full models at about $\frac{V}{\sqrt{L}} = 1$, while for fine models this hump is imperceptible or shows itself only as an unfair portion of the curve and the first important hump is at about $\frac{V}{\sqrt{L}} = 1.4$ to 1.5.

For quite full models, especially those with parallel middle body, the hump for $\frac{V}{\sqrt{L}} = .8$ is often important, and for such models the hump for $\frac{V}{\sqrt{L}} = .67$ to $.7$ is frequently detected though not of importance.

The values of $\frac{V}{\sqrt{L}}$ above refer to the centers of the humps or the points where the percentage increase of resistance above an average curve is a maximum. Of course, the departures from the average begin and end some distance before and beyond the hump centers.

Fig. 66 shows graphically the relations between speed of ship, length in feet and values of $\frac{V}{\sqrt{L}}$. By using a varying scale for length, the abscissæ being proportional to $\sqrt{\text{length}}$, the contours of $\dfrac{V}{\sqrt{L}}$ are straight lines. By shading the regions corresponding to humps and leaving clear those corresponding to hollows the relative locations of humps and hollows are indicated. It will be observed that the two lower humps of Fig. 66 are indicated at slightly lower values of $\dfrac{V}{\sqrt{L}}$ than in Fig. 65. This is because Fig. 65 is for a constant value of m, namely 1.15, while in practice we find for the lowest hump $m = 1.00$ very nearly, and for the next $m = 1.08$ or so. For the region from $\dfrac{V}{\sqrt{L}} = .9$ to $\dfrac{V}{\sqrt{L}} = 1.2$, embracing a hump and a hollow, $m = 1.15$ very nearly while beyond this speed m is somewhat greater on the average.

It might seem at first sight very important to adopt such length for a desired speed as to be sure of landing in a hollow rather than on a hump, but, though this point should always be considered, in comparatively few cases is it a matter of serious practical importance. In most cases it is desirable to adopt proportions and form such that the humps and hollows up to the speed attained are not prominent, so there is no material saving to be had by landing in a hollow rather than on a hump.

4. Curves of Residuary Resistance and of Coefficients. — Having discussed generally the characteristics of wave-making resistance as indicated by the formula

$$R_w = V^4 \left(A^2 + B^2 + 2\, kAB \cos \frac{m}{c^2} 646° \right),$$

it is well to consider some concrete examples.

Fig. 67 shows curves of residuary resistance determined from model experiments for ten 400-foot ships without appendages. The residuary resistance is practically all wave-making. The proportions, etc., are tabulated on the figure.

It is seen that there are five displacements in all, there being two vessels of each displacement differing in midship area or longitudinal coefficient. All vessels were derived originally from the same parent lines, so the variations of resistance are essentially due to variations of dimensions and of longitudinal coefficient.

The curves of Fig. 67 are not very encouraging to the development of an approximate formula for wave resistance.

For instance the variation with longitudinal coefficient is a very difficult feature. The models of .64 longitudinal coefficient all show pronounced humps at about 21 knots, while their mates of .56 longitudinal coefficient show no hump there. But at 25 knots or so the wave resistances for the two coefficients come together again, and for higher speeds the models of .64 coefficient have the smaller resistances. At 30 knots or so there is a second hump which shows for both the full and the fine coefficients.

Resistance curves are frequently analyzed by assuming them of the form $R = AV^n$ and determining suitable values of n, the power of the speed according to which the resistance is varying, and of a, the corresponding coefficient. The curves of Fig. 67 are analyzed in this way without much trouble by plotting them upon logarithmic section paper. For a curve so plotted the exponent n at a point is proportional to the inclination of the curve.

Fig. 68 shows curves of the exponent n for the 10 curves of wave resistance of Fig. 67. It is seen that the variations of n are enormous. As to a in the formula $R = aV^n$ the values corresponding to the curves of n in Fig. 68 vary too rapidly and radically to be adequately represented graphically.

Suppose now we attempt a slightly different analysis. We have deduced a qualitative formula for wave resistance as follows:

$$R_w = V^4 \left(A^2 + B^2 + 2\, kAB \, \cos \frac{m}{c^2} 646° \right).$$

Then curves of $\dfrac{R_w}{V^4}$ will also be curves of

$$A^2 + B^2 + 2\, kAB \, \cos \frac{m}{c^2} 646°$$

and might be expected not to vary very much. Fig. 69 shows curves of $\dfrac{R_w}{V^4}$ for the 10 curves of Fig. 67, the residuary resistance R_r being taken as identical with R_w. It is seen that up to 18 knots or so these curves are reasonably constant. Here they begin to rise. For the full coefficients there is a maximum at 21 knots or so, a minimum at 23 to 24 knots and a second maximum at 29 to 30 knots. For the fine coefficients there is only one pronounced maximum at 29 to 30 knots.

It is evident from Fig. 69 that the curves of $\frac{R_w}{V^4}$ are somewhat systematic in their variations and that it might be possible to formulate values of A, B, k and m such that in a given case we could determine R_w with reasonable approximation from the basic formula

$$R_w = V^4\left(A^2 + B^2 + 2\,kAB\,\cos\frac{m}{c^2}\,646°\right).$$

It is equally evident that the formulæ for A, B and c involved would be difficult and complicated. It will be shown later that by graphic methods the residuary resistance in a given case can be readily approximated and hence the task of devising approximate formulæ need not be undertaken.

It is interesting to note for ships 1 to 4 the relative reduction in wave resistance beyond 30 knots.

The reason will be made clear upon reference to Fig. 66. It is seen that for a 400-foot ship the last hump occurs at about 30 knots. In this condition the wave length corresponding to the speed is somewhat greater than the length of the ship, so that the second crest of the bow wave is superposed upon the first crest of the stern wave. Hence the hump. At a speed of about 40 knots there would be a final hollow corresponding to the conditions when the first hollow of the bow wave is superposed upon the first crest of the stern wave. This is the main cause of the apparent relative falling off of wave resistance in Figs. 68 and 69 between 30 and 40 knots.

Fig. 66 would indicate that some distance beyond 40 knots the wave resistance of these 400-foot ships would again begin to increase relatively, but there is some reason to believe that at excessive speeds — say 120 knots for the 400-foot ships — the wave resistance would be decreased by the bodily rise of the ship, which would begin to approach the condition of a skipping stone and tend to glide along the surface. Of course, the speed of 120 knots is unattainable by any 400-foot ship at present, but it corresponds to 36 knots for a 36-foot boat, which is not very far beyond the speed-launch results now attainable. Consideration of such extreme cases is, however, beyond the scope of this work.

11. Air Resistance

The above water portions of a ship may be regarded as immersed in the air, and air, like water, offers resistance to the motion of a body surrounded by it. Air is, roughly, only from one-ninth to one-eighth of one per cent of the weight of water, the actual weight depending on the pressure and temperature, and air resistances compared with those of water are, roughly, as the relative densities. But air resistance is by no means always negligible. Sailing vessels are driven by the resistance of sails to the motion of air past them, and any one who has attempted to stand on the deck of a vessel exposed to a gale of wind will admit that a strong head wind opposes a good deal of resistance to a vessel with even a moderate amount of top-hamper.

1. **Zahm's Experiments upon Air Friction.** — Air resistance can be separated into two classes — frictional and eddy resistance. Careful investigations of the friction of air upon plane surfaces have been made by Prof. A. F. Zahm, of Washington, who in a paper of February 27, 1904, before the Philosophical Society of Washington (Bulletin, Vol. XIV, pp. 247–276) has given experimental results for air friction upon thin planes somewhat similar to those tried in water by Froude.

Prof. Zahm's air planes were $25\frac{1}{2}$ inches wide, one inch thick, and of varying lengths up to 16 feet. While rather smaller than Froude's planes, they were tried up to a high air velocity of 25 statute miles per hour, or $21\frac{3}{4}$ knots.

Prof. Zahm summarizes his most important conclusions upon the subject of air resistance as follows:

1. The total resistance of all bodies of fixed size, shape and aspect is expressed by an equation of the form $R = av^n$, R being the resistance, v the wind speed, a and n numerical constants.

2. For smooth planes of constant length and variable speed, the tangential resistance may be written $R = bv^{1.85}$.

3. For smooth planes of variable length l and constant width and speed the friction is $R = cl^{0.93}$.

4. All even surfaces have approximately the same coefficient of skin friction.

5. Uneven surfaces have a greater coefficient of skin friction, and the resistance increases approximately as the square of the velocity.

These conclusions as to air friction are in striking agreement with those deduced by Froude for surface friction in water.

The coefficients given by Zahm are readily reduced for speeds in knots instead of feet per second or statute miles per hour.

Upon doing this, if R denote frictional air resistance in pounds, A denote whole area of surface in square feet, l denote length of surface in feet and V denote speed through the air in knots, we have
$$R = .0000122 \, l^{.93} A V^{1.85}.$$

It should be remembered that this formula is based upon experiments with planes no longer than 16 feet tested up to speeds of 25 statute miles per hour. So, while it may be used with confidence for short planes up to any velocity reached by ships, it must be regarded as only a fair approximation for long surfaces. Fortunately for the purpose of the naval architect a fair approximation to frictional air resistance is all that he ever need know in practice. It is very seldom indeed that he will need to take any account of it at all.

For convenience in calculation Table VIII gives values of $V^{1.85}$ and of $\frac{1}{l^{.07}}$. We have $l^{.93} = \frac{l}{l^{.07}}$, and hence can readily obtain $l^{.93}$ if we know $\frac{1}{l^{.07}}$. A table of $l^{.93}$ would not admit of easy interpolation, while $\frac{1}{l^{.07}}$, which varies comparatively slowly, lends itself to interpolation.

Comparing the results of his experiments on air friction with those of Froude on water friction, Zahm states:

" With a varnished board 2 feet long, moving 10 feet a second, the ratio of our coefficients of friction for air and water is 1.08

times the ratio of the densities of those media under the conditions of the experiments."

Froude, however, found that the coefficient of friction fell off more rapidly with length than as $l^{-0.07}$, so that for longer planes the above ratio is greater than 1.08 times the ratio of densities. Thus, for 20-foot planes the ratio of coefficients would be some $1\frac{1}{2}$ times the density ratio, that is, the friction in air would be $1\frac{1}{2}$ times that deduced from water friction by dividing it by the density ratio.

Zahm states that in his experiments " no effort was made to determine the relation between the density and skin friction of the air, partly for want of time, partly because, with the apparatus in hand, too great changes of density would be needed to reveal such relation accurately. Doubtless the friction increases with the density."

It appears probable that we may assume Zahm's formula for frictional resistance of air to apply to air at 60° F. and a barometer pressure of 30 inches.

2. Eddy Resistance in Air. Results of Experiments with Planes. — While the frictional resistance of air is of importance in connection with flying machines, for ships the most important air resistance is the eddy resistance.

The eddy resistance of air seems to follow the same general laws as the eddy resistance of water. Within the limits of the speed attained by the wind, say up to 100 miles per hour, it varies for a given plane as the square of the speed. Observations made under the direction of Sir Benjamin Baker during the construction of the Forth Bridge indicated that small planes exposed to the wind offered greater resistance per square foot than larger planes exposed to the same wind. M. Eiffel found for planes not over 1 meter square falling through still air that the larger planes showed slightly greater resistance per square foot.

For rectangular planes the resistance varies somewhat with the ratio of the sides, a long narrow plane offering greater resistance than a square of the same area.

For our purposes it is not necessary to consider closely these minutiæ, and it will suffice to use an average coefficient and express the resistance in pounds of a plane of area A in square feet moving normally through the air with velocity V knots by a single formula $R = CAV^2$.

The values of the coefficient C which have been obtained by various experimenters vary a good deal. The more recent experimenters seem to obtain the lower values, but coefficients obtained by experimenters within the last 30 to 40 years range from .0035 to .005 about.

In England, Stanton, with very small planes exposed to a current of air through a large pipe or box, has obtained a coefficient of .0036. Dines with rather small planes on a whirling arm has obtained .00384. Mr. William Froude with good-sized planes moving through still air at rather low velocities obtained .0048. In America, Langley, by whirling-arm methods, obtained somewhat variable coefficients averaging about .0047. In France, quite recently, M. Eiffel, with planes up to 10 square feet or more in area, falling through still air, conducted very careful and elaborate experiments and obtained a coefficient of .004. (See " Recherches Expérimentales sur la Résistance de l'Air Exécutées à la Tour Eiffel par G. Eiffel." This was published in 1907.)

All things considered, in the light of our present experimental knowledge on the subject it appears reasonable to adopt the coefficient .0043 as suitable for practical use. Then our formula for the resistance in pounds of a plane moving normally to itself is $R = .0043 \, AV^2$, where A is area in square feet and V is speed in knots. For speed in statute miles the coefficient above should be divided by 1.326; for speed in feet per second by 2.853.

When it comes to the normal pressure on an inclined plane moving through the air the results obtained by experimenters are somewhat peculiar. For square planes and rectangular planes whose sides are not too dissimilar the normal pressure increases rapidly from zero at zero inclination up to an inclination of 30 degrees or so. At this inclination the normal pressure is nearly the same as at 90° inclination, and from 30° to 90° inclination the normal pressure, while varying somewhat irregularly, does not change much.

The simplest formula is that of M. Eiffel. For inclined planes he proposes to take the normal pressure as constant from 30°

to 90°, and from 0° to 30° to take it as varying linearly. The Eiffel formula is a sufficiently close approximation for practical use.

The formula, then, for practical use expressing the normal pressure in pounds P_n on an inclined plane moving through the air at an angle of θ degrees will be

From 0 to 30°, $P_n = .0043 \dfrac{\theta}{30} A V^2$;

above 30°, $P_n = .0043\, A V^2$, where A is area in square feet and V is speed in knots.

The normal pressure is, of course, different from the resistance in the direction of motion, which is $P_n \sin \theta$, or the component of P_n parallel to the direction of motion.

3. Determination of Air Resistance of Ships. — There is no practical method recognized at present for determining the air resistance of a ship. Mr. William Froude made some experimental investigations of the matter about 1874, in connection with the *Greyhound*, a vessel 172.5 feet × 32.2 feet × 13 feet draught, of about 1000 tons displacement. The vessel was tried without masts or rigging. He concluded that in this condition at 10 knots, the air resistance of the *Greyhound* was nearly 150 pounds, or about 1½ per cent of the water resistance.

For steamers without large upper works, the air resistance, when the air is still, is, without doubt, too small as a rule to require much consideration. With a strong head wind the air resistance is, of course, very much increased, but under such conditions the increase of water resistance due to the head sea is probably in most cases far greater than the air resistance. In cases where air resistance is important, it can be investigated by exposing a model with the upper works complete to a current of air of known speed. The law of the square applies, and it will be possible to determine the air resistance of the model at the actual speed, not the corresponding speed of the ship. Then the air resistance of the full-sized ship, being practically all eddy resistance, may be estimated by multiplying the resistance of the model at the speed of the ship by the square of the ratio between the linear dimensions of the ship and the model.

For a rough approximation, we may take the area of the portion of the ship above water projected on a thwartship plane and assume that the air resistance is that due to a plane of this area — denoted by A — advancing normally through the air, using the formula already given for the resistance of a plane. This would give us —

Air Resistance in pounds = $.0043\ AV^2$, where V is speed through the air in knots and A is area of upper works projected on a thwartship plane.

12. Model Experiment Methods

In view of the very large use now made of model-basin experiments there will be given a brief description of the methods used in deducing from the model experimental results the resistance or effective horse-power of the full-sized ship.

At a model tank or basin there are facilities for making to scale models of ships representing accurately the under-water hulls and a sufficient amount of the above-water hulls. Most model basins work with models from 10 to 12 feet long. Some use models as long as 20 feet. A complete model can be towed through the still water of the basin, the speed and corresponding resistance being measured for a number of speeds covering the range desired.

1. Treatment of Model Results. — By plotting each resistance as an ordinate above its speed as an abscissa we obtain a number of spots through which a fair average curve is drawn, giving the total resistance of the model. Fig. 70 shows for an actual model a number of experimental spots and the resistance curve drawn through them. When reducing the results the first step in practically all cases is to determine the estimated frictional resistance of the model.

The wetted surface of the model has been calculated and we have recorded from experiments with planes the length of the model, the resistance of a square foot of surface for each tenth of a knot extending up to any speed to which a model is likely to be tested. Fig. 70 shows a curve of r_f or frictional resistance of model, its ordinates having been determined for various speeds by multiplying the model surface by the resistance of one square foot.

2. Deduction of Ship Resistance, Using Model Results.

— For the most common case the model represents some full-sized ship — actual or designed — and we wish by the aid of the model results to determine a curve of estimated effective horse-power for the full-sized ship.

Table IX herewith gives the calculations for the *Yorktown*, for whose model the resistance curve is given in Fig. 70. The object of much of the form is obvious. The "Mean Immersed Length," L, of the ship is usually the length on the load water line. For models of peculiar profiles there is a correction applied by judgment, the object being to obtain the average immersed length. The mean immersed length of the model is usually made 20 feet at the United States Model Basin, though moderate departures from this length are made when desirable for any reason. Also, as it is difficult to get satisfactory observations above a speed of 17 knots of model, it is necessary to make models shorter than 20 feet if the maximum corresponding speed would be over 17 knots for a 20-foot model.

The model is so weighted that if it is exact it will float in the fresh water of the basin at exactly the corresponding water line of the ship in salt water. Hence the ratio at corresponding speeds of resistances which follow Froude's Law is not $\left(\frac{L}{l}\right)^3$ but $\frac{36}{35}\left(\frac{L}{l}\right)^3$, the factor $\frac{36}{35}$ being introduced on account of the passage from fresh water to salt water.

Coming now to the tabular form, there are entered in the first column values of v or the speed of the model in knots, and in the second column corresponding values of r or the total resistance of the model in pounds as taken from the curve in Fig. 70. In the third column is entered r_f or the frictional resistance of the model — calculated as already described. In the fourth column we enter the residuary resistance, r_r, which is equal to $r - r_f$. It is this resistance to which Froude's Law applies, and we wish to deduce from it in the shortest and simplest manner the corresponding residuary effective horse-power. While r_r is mostly Wave Resistance, it includes the Eddy Resistance and Air Resistance of the model. Both are taken as following the Law of Comparison.

Now for the full-sized ship the residuary resistance in pounds at corresponding speed is $r_r \times \frac{36}{35}\left(\frac{L}{l}\right)^3 = R_r$, say. The speed of ship, V, corresponding to a speed of model, v, is $v\sqrt{\frac{L}{l}}$, and the effective horse-power absorbed by R_r is $R_r \times .0030707\, V$. Then, if the residuary effective horse-power for the full-sized ship is denoted by EHP_r we have

$$EHP_r = R_r \times .0030707\, V = r_r \frac{36}{35}\left(\frac{L}{l}\right)^3 \times .0030707\, v\sqrt{\frac{L}{l}}$$

$$= r_r\, v\, \frac{36}{35}\left(\frac{L}{l}\right)^3 \times .0030707\, \sqrt{\frac{L}{l}}.$$

We denote by a the quantity $\frac{36}{35}\left(\frac{L}{l}\right)^3 .0030707 \sqrt{\frac{L}{l}}$ and calculate it once for all, as indicated in the heading. Then in the fifth column of the table we enter av and in the sixth column EHP_r, which is simply r_r multiplied in each case by av. In the ninth column we enter V, the corresponding speed for the ship, obtained by multiplying each value of v by $\sqrt{\frac{L}{l}}$. We have now for a number of values of V the values of EHP_r or residuary effective horse-power.

We need to determine the frictional portion of the effective horse-power. This is denoted by E_f or EHP_f. To determine frictional resistance we take from Table VI of Tideman's Constants the coefficient of friction appropriate to the length of the vessel and the nature of bottom. The area of wetted surface has been calculated.

We have seen that frictional resistance in pounds $= R_f =$ wetted surface $\times$ frictional coefficient $\times V^{1.83}$
and $E_f = .0030707\, R_f \times V$
$\quad\quad = .0030707 \times$ wetted surface $\times$ frictional coefficient $\times V^{2.83}$.
Taking from Table VII the values of $V^{2.83}$ we readily determine and enter in column 11 the values of EHP_f. These values are plotted as in Fig. 71 and a fair curve run through. Then from

this curve for the values of V_{cor} in column 9 we take off the values of EHP_f and enter them in column 7. Column 8, which is the sum of columns 6 and 7, gives the values of the total EHP, which, spotted in Fig. 71 over the values of V_{cor}, enables us to draw the final curve of E.H.P. for the condition of the ship defined in the heading of the table.

From this curve it is possible to fill in column 12, which gives the values of E.H.P. corresponding to the even values of V in column 10. Column 12 is, however, seldom needed.

3. Residuary Resistance Plotted for Analysis. — When we are dealing with an actual ship or design it is generally desirable to deduce from the model results the final E.H.P. curve as soon as possible. When, however, it is a question of analysis of residuary resistance it is desirable to express it in a slightly different form. A very convenient and instructive method is to use the values of $\dfrac{V}{\sqrt{L}}$ as abscissæ and of Resistance ÷ Displacement as ordinates.

For convenience the value of Resistance ÷ Displacement is expressed as Resistance in Pounds per Ton of Displacement. Fig. 72 shows the curve of Residuary Resistance in Pounds per Ton plotted on $V \div \sqrt{L}$ for the model to which Figs. 70 and 71 refer. Fig. 72 is applicable to any size, and it is this elimination of the size feature which renders this method of plotting of value for purposes of analysis.

13. Factors Affecting Resistance

The problem of resistance in its most general form involves too many variables to be capable of experimental solution. For a vessel of given displacement and speed the resistance varies with variations of (1) The dimensions, (2) The shapes of water lines and sections. For a vessel of given displacement we may have an infinite number of variations of dimensions and shape, so even if we could deduce the resistance of a vessel with mathematical accuracy from model experiments, it would be a formidable undertaking to investigate all admissible or likely variations of dimensions and shape for but a single vessel of a fixed displacement.

1. **Derivation of Models from Parent Lines.** — If, however, we adopt a single definite shape or set of parent lines, deducing all models from these lines by variations of dimensions and coefficients of fineness, the problem is enormously simplified. By testing a practicable number of models we can determine, not for one displacement only, but for any displacement within a certain range and for any dimensions and fineness likely in practice, the approximate resistance at any practicable speed.

In connection with fineness the expression "longitudinal coefficient" will be used to denote the ratio between the volume of displacement of a vessel and the volume of a cylinder of section the same as the submerged midship section and of length the same as the length of the vessel — preferably the mean immersed length. This coefficient is sometimes called the "cylindrical coefficient" and very commonly the "prismatic coefficient." While cylindrical coefficient is descriptive and correct, it is thought that the designation "longitudinal coefficient" is preferable as emphasizing the fact that this coefficient measures and expresses the fineness of the vessel in a longitudinal direction. The expression "prismatic coefficient" is slightly in error, since strictly speaking the section of any prism is bounded by a straight-sided polygon and not by a curve.

Given a set of parent lines, the deduction from them of lines of the same coefficients but of different proportions or relative values of length, beam and draught, is a simple matter. If length alone is changed, we need only change the spacing of stations in proportion to the change of length. If draught alone is changed, we need change only in a corresponding way the spacing of water lines. If beam alone is changed, we need change only the ordinates of water lines.

Since the changes caused by change of length, beam and draught are independent we may simultaneously change all three, if we wish, without difficulty.

Suppose, however, we wish to keep dimensions unchanged and make changes in shape and fullness. We cannot change the midship section without departing from the parent lines, but we can change in a comparatively simple manner the longitudinal coefficient or curve of sectional areas. Thus in Fig. 73, suppose the curve numbered 1 is the curve of sectional areas for the parent model and the curve numbered 2 the desired curve of sectional areas. Through E, the point on curve 2 corresponding to the station AB, draw EF horizontally to meet curve 1 at F. Through F draw CD, then the proper section at AB of the derived form is the section at CD of the parent form. Having the two curves of sectional area and the half-breadth plan of the parent form, any desired section of the derived form can be determined without difficulty.

From a single parent form then, we can derive forms covering all needed variations of displacement, of proportions and of fineness as expressed by "longitudinal coefficient." By contour curves from the results of a number of models derived from one parent form we can deduce diagrams enabling us to ascertain the resistance at any speed of any vessel upon the lines of the parent form. This applies, of course, to residuary resistance only, since the frictional resistance can always be estimated without model results or experiments in the manner already indicated.

2. Classification of Factors Affecting Resistance. — It would require experiments with models derived from an infinite number of parent forms to trace the effect of all possible variations of shape, but if we can determine the major factors affecting resistance and their approximate effect we need seldom concern ourselves with the minor factors.

While it is necessary to be cautious in laying down from past experience a hard and fast line of demarcation between the major and minor factors of resistance, since novel developments in the future may convert one into the other, yet so far as can be judged from trials at the United States Model Basin of over a thousand models we appear warranted in drawing some conclusions as to the principal factors affecting the resistance of ships not of abnormal form and the relative importance of these factors. We need consider only frictional and wave-making or residuary resistance.

Given the displacement, speed and frictional quality of the surface, the only other factor of importance as regards frictional resistance is the length. The greater the length for a given displacement the greater the frictional resistance. This because frictional resistance is proportional to surface or $\sqrt{DL}$.

As regards residuary resistance for a given displacement, the principal factors arranged in their usual order of importance are as follows:

1. The length.
2. The area of midship section or, conversely, the longitudinal coefficient.
3. The ratio between beam and draught.
4. The shape of midship section or midship section coefficient.
5. The details of shape toward the extremities.

It is seen that factors 1, 2 and 3 can be investigated from a single parent form. The complete investigation of factors 4 and 5 would require investigations involving a very large number of parent forms. Fortunately, however, these factors are those of least importance.

3. Details of Shape Forward and Aft. — In placing factor 5 as of small importance, it should be understood that this is the case only as regards the variations found in good practice. If abnormal shapes for the extremities are adopted, abnormal resistance is liable to follow. The dictum of William Froude many years ago appears to be still our best guide. He stated that, broadly speaking, it was desirable to make the bow sections of U shape and the stern sections of V shape. This amounts to saying that at the bow it is advisable to put the displacement well below water and make the water line narrow, and at the stern it is advisable to bring the displacement up towards the surface and make the water line broad. Carried to an extreme, this would give us hollow water lines at the bow and the broad flat stern of the torpedo boat type. As a matter of fact, model basin experiments appear to indicate that for smooth water, up to quite a high speed, this type of model is about the fastest. For extreme speeds, even in smooth water, hollow bowlines are seldom adopted, but there is not sufficient experience in this connection to say positively that they are or are not desirable from the point of view of speed alone.

In this connection it may be pointed out that experiments show a ram bow of bulbous type to be favorable to speed, even apart from the fact that the ram bow usually involves a slight increase in effective length. This is simply because the ram bow, which is the extreme case of the U bow, is much fuller below water than at the water line.

The excess pressures set up around the ram being well below the surface are more absorbed in pumping the water aft, where it is needed, and less absorbed in raising the surface and producing waves than if the same displacement were brought close to the surface.

There appears to be a reasonable explanation of the advantages as regards resistance of the broad flat stern. In wake of the center of length, the water is flowing aft to fill up the space being left by the stern, the greatest velocity of the water being under the bottom. As the vessel passes, the water flows aft and up, losing velocity all the while and increasing in pressure.

With a U stern there is little to check the upward component of the velocity which is absorbed in raising a wave aft. With the broad flat stern against which the water impinges, as it were, more or less of the upward velocity is absorbed by pressure against the stern, which will have a forward component, the result being a closer approach to perfect stream motion and less wave disturbance.

While the broad flat stern is slightly superior as regards residuary resistance in smooth water, it is apt to have unnecessary wetted surface and is objectionable from a structural and seagoing point of view. With model basin facilities it is generally possible to determine upon a stern of V type which is almost as good as the broad flat type as regards resistance, and distinctly preferable to it from a structural and sea-going point of view.

In connection with the details of shape forward and aft the effect of change of trim upon resistance may be considered, since the principal effect of change of trim is to modify the shape towards the extremities.

Any change of trim, no matter how small, necessarily produces some effect upon resistance, and there are many sea-going people who ascribe great virtue to some particular trim and great influence upon resistance to change of trim, generally considering trim by the stern as advantageous for speed.

Trim by the stern has some advantages in that it generally improves the steering of the ship or its steadiness on a course, and in rough weather it is generally advantageous to secure greater immersion of the screws and more freeboard forward; but as regards resistance in smooth water changes of trim occurring in practice generally produce changes of resistance of little or no importance.

In 1871 Mr. William Froude investigated the effect of trim upon the resistance of the *Greyhound*, a vessel 172 feet long and towed at displacements from 938 to 1161 tons and at trims varying from 1.5 feet by the head to 4.5 feet by the stern. The maximum speed at which the vessel was towed was about 12 knots. These experiments showed that for the *Greyhound* trim by the head was beneficial at low speeds, below 8 knots, and trim by the stern was beneficial at the upper speeds, above 9 knots. The differences, however, were comparatively small for quite large changes of trim. Mr. Froude's conclusion from these full-sized towing experiments was, "As dependent on differences of trim, the resistance does not change largely; indeed, at speeds between 8 and 10 knots it scarcely changes appreciably, even under the maximum differences of trim." The results from the *Greyhound* were corroborated by model experiments which agreed quite well with the full-sized results, and since these classical experiments of Mr. Froude, model experiments investigating this question have been repeatedly made.

Many experiments made at the United States Model Basin appear to indicate that, broadly speaking, for the majority of actual vessels at full speed a slight trim by the stern is beneficial, but that in the vast majority of cases the benefit is too small to be of practical importance. With a well-balanced design, the fineness forward and aft being properly distributed, the effect upon resistance of change of trim is practically nil.

4. Shape of Midship Section. — Let us now consider the influence upon resistance of midship section fullness or the midship section coefficient. Figs. 50 to 54 show body plans of five models, all having the same length, the same displacement — 3000 pounds — the same curve of sectional areas, the same area of midship section and practically the same load water line. Figs. 55 to 59 show similarly body plans of five 1000 pound models.

Each group of five models has midship section coefficients varying from .7 to 1.1, the models with fine midship section coefficients having greater values of B and H since the actual midship section areas are the same for all models of a group. The ratio $B \div H$ for all ten models is 2.92. The models are of moderately fine type, the longitudinal coefficient being .56 for all ten. Fig. 74 shows curves of residuary resistance in pounds per ton for the five 3000 pound models and Fig. 75 shows similarly the resistances of the five 1000 pound models.

It is seen that while the models with full midship section coefficients drive a little easier up to $V \div \sqrt{L} = 1.1$ to 1.2 and the models with fine coefficients have a shade the best of it at higher speeds, the differences for such variations of fullness as are found in practice are remarkably small. The results given above are taken from a paper by the author before the Society of Naval Architects and Marine Engineers in November, 1908, on "The Influence of Midship Section Shape upon the Resistance of Ships." This paper contained many other results similar to those given, and its conclusion was that "for vessels of usual types and of speeds in knots no greater than twice the square root of the length in feet, the naval architect may vary widely midship section fullness without material beneficial or prejudicial effect upon speed." Of course, it follows that the minor variations in shape of midship section that can be made in practice without changing fullness have practically no effect upon resistance.

It should be most carefully borne in mind that the above applies to the shape and coefficient of a midship section of a given area, not to the area of the section.

5. Ratio between Beam and Draught. — Consider now the effect of the ratio between beam and draught. Figure 76 shows curves of E.H.P. as determined by model experiment for 6 vessels, all derived from the lines of the U. S. S. *Yorktown* but varying in proportions of beam and draught from a very broad shallow model to a very narrow deep one.

It is seen that the broader and shallower the model the greater the resistance. This result is typical and confirmed by many other experiments at the United States Model Basin. It may at first sight seem opposed to many cases of experience where beamy models proved easy to drive. But in these cases it will be found that the increase of beam carried with it increase of area of midship section. Had beam been increased and draught decreased in proportion, the area of midship section remaining unchanged, the results would have been different.

However, the variations of resistance with variations of the ratio of beam to draught are not very great as a rule.

6. Longitudinal Coefficient or Midship Section Area. — Take up now the effect upon resistance of the variation of midship section area or longitudinal coefficient. This is a factor of prime importance in some cases and quite secondary in others. Thus, Fig. 67 shows curves of residuary resistance for five pairs of 400-foot ships, each pair having the same displacement and derived from the same parent lines but differing in midship section area or longitudinal coefficient. It is seen that at 21 knots No. 10 with .64 longitudinal coefficient has 2.3 times the residuary resistance of its mate No. 9 with .56 longitudinal coefficient. But at $24\frac{1}{2}$ knots they have the same residuary resistance.

Again, No. 4 of .64 coefficient at 21 knots has nearly twice the residuary resistance of No. 3 of .56 coefficient. At $25\frac{1}{2}$ knots they have the same residuary resistance and at higher speeds No. 4 has the best of it, having but .9 of the residuary resistance of No. 3 at 35 knots. These results, which are thoroughly typical, are susceptible of a very simple qualitative explanation. A small longitudinal coefficient means large area of midship section and fine ends. A large longitudinal coefficient means small area of midship section and full ends. At moderate speed the ends do the bulk of the wave making and the fine ends make much less wave disturbance than the full ends. Hence the enormous advantage of the fine ends at 21 knots in Fig. 67. But at high speeds the whole body of the ship takes part in the wave making and the smaller the midship section the less the wave making. It follows that for a ship of given dimensions, displacement, type of form and speed there is an optimum longitudinal coefficient or area of midship section. Data will be given later by which this can be determined with close approximation.

7. Effect of Length. — There remains finally to consider the factor which, broadly speaking, has more influence upon residuary resistance than any other. This is the length. We have seen that for a given displacement the greater the length the greater the frictional resistance — it varying as $\sqrt{L}$. Residuary resistance, on the contrary, always falls off as length increases, though not according to any simple law. Fig. 77 shows curves or residuary resistance of five vessels, all of 5120 tons, derived from the same parent lines and having the lengths given. Of course the longer vessels have beam and draught decreased in the same ratio sufficiently to keep the displacement constant. Fig. 77 illustrates very clearly the enormous influence of length upon residuary resistance. Since frictional resistance increases and residuary resistance decreases with length, it is reasonable to suppose that for a given displacement and speed there will be a length for which the total resistance will be a minimum. There is such a length, but in the vicinity of the minimum the increase of resistance with decrease of length is slow, and since length in a ship is usually undesirable from every point of view except that of speed, ships should be made of less length than the length for minimum resistance. For men-of-war particularly it is good policy to shorten the ship, put in slightly heavier machinery and accept the increased coal consumption upon the rare occasions when steaming at full speed, rather than to lengthen the ship, carry greater weight of hull and armor necessitated thereby, and consume more coal at ordinary cruising speeds.

14. Practical Coefficients and Constants for Ship Resistance

1. Primary Variables Used. — The first thing to do when we wish to establish methods for the determination of ship resistance is to fix the primary variables to be used. In a given case we may have dimensions, displacement, etc., all fixed, and need to determine the resistance at a given speed, or we may wish to determine dimensions to bring resistance below a certain amount, or the problem may present other aspects. The primary variables adopted should enable the data available to be applied simply and directly to the problems arising.

It is convenient to express resistance as a fraction of displacement, and a suitable measure is the resistance in pounds per ton of displacement. Then a resistance of one pound per ton of displacement means a resistance which is $\frac{1}{2240}$ of the displacement. At corresponding speeds for similar models, resistances which follow Froude's Law are proportional to displacement, and hence the pounds per ton are constant.

Speed is conveniently expressed not directly but in terms of $\frac{V}{\sqrt{L}}$, the speed length ratio or speed length coefficient. For similar models at corresponding speeds $\frac{V}{\sqrt{L}}$ is constant.

When it comes to size we need a variable which does not change for similar models whatever the displacement. Since the displacement varies as the cube of linear dimensions, such a quantity would be Displacement ÷ (any quantity proportional to the cube of linear dimensions). As length is much more important in connection with resistance than beam or draught, a suitable quantity would be $\frac{D}{L^3}$. This would usually be a very small fraction, however, and it is desirable to use a function which in practical cases assumes numerical values convenient for consideration and comparison. Such a function is $\dfrac{D}{\left(\dfrac{L}{100}\right)^3}$, called the displacement length ratio or displacement length coefficient. It is the displacement in tons of a vessel similar to the one under consideration and 100 feet long.

2. Skin Resistance Determination. — It is necessary to consider separately the two elements of resistance, Skin Resistance and Residuary Resistance.

The former is the greater in most practical cases and its independent calculation is very simple. We have seen that the formula for Skin Resistance is $R_f = fSV^{1.83}$, where f is coefficient of friction from Tideman or Froude, S is wetted surface and V is speed in knots. For a complete design S may be accurately calculated. For a preliminary design it may be closely estimated from the formula $S = c\sqrt{DL}$, where c is the wetted surface coefficient and may be taken from Fig. 41.

If we were concerned with Skin Resistance only, it would probably be the best plan always to determine E.H.P.$_f$ by formula as was done when calculating the E.H.P.$_f$ of a full-sized ship from the results of model experiments. But it is necessary to use a more complicated system of variables in order to handle Residuary Resistance, so it is desirable to express R_f in the same variables.

We have seen that $R_f = fSV^{1.83}$ and $S = c\sqrt{DL}$. Hence $R_f = fc\sqrt{DL}V^{1.83}$.

Write $y = \dfrac{D}{\left(\dfrac{L}{100}\right)^3}$. Then $y = \dfrac{1000000\,D}{L^3}$ or $D = \dfrac{yL^3}{1000000}$.

Also write $x = \dfrac{V}{\sqrt{L}}$. Then $V = x\sqrt{L}$ $V^{1.83} = x^{1.83}L^{0.915}$.

Then
$$\frac{R_f}{D} = fc\frac{\sqrt{DL}V^{1.83}}{D} = fc\sqrt{\frac{L}{D}}V^{1.83} = fc\sqrt{\frac{1000000\,L}{yL^3}}x^{1.83}L^{0.915}.$$

Whence finally
$$\frac{R_f}{D} = 1000\,fc\frac{x^{1.83}}{y^{\frac{1}{2}}} \times \frac{1}{L^{.085}},$$

or
$$\frac{R_f}{D} = \frac{1000\,fc}{L^{.085}}\frac{\left(\dfrac{V}{\sqrt{L}}\right)^{1.83}}{\left[\dfrac{D}{\left(\dfrac{L}{100}\right)^3}\right]^{\frac{1}{2}}}.$$

In the above f varies slightly with length, $L^{.085}$ varies slowly with length, and c is an almost constant coefficient.

Evidently then for a given length and value of c we can plot contours of $\dfrac{R_f}{D}$ on $\dfrac{V}{\sqrt{L}}$ and $\dfrac{D}{\left(\dfrac{L}{100}\right)^3}$ as primary variables. Fig. 78 shows such contours for a length of 500 feet, the value of f being taken from Table VI of Tideman's constants. But $\dfrac{f}{L^{.085}}$ does not vary very rapidly with length and it varies with length only.

So Fig. 78 can be applied to all lengths and values of c by the use of simple correction factors. The correction factors for length are given on the scale beside the figure to the right. In Fig. 78 the standard value assumed for c is 15.4. If we are dealing with a vessel for which we know that c is 16.0 for instance, it is obvious that we should multiply the values of $\frac{R_f}{D}$ from Fig. 78 by $\frac{16.0}{15.4}$.

3. Residuary Resistance from Standard Series. — Take up now the question of Residuary Resistance. Here we are driven to the use of model results.

Fig. 79 shows the lines used for a series of models which may be called the Standard Series.

Fig. 79 shows a model having a longitudinal coefficient of .5554, a midship section coefficient of .926 and a displacement length ratio of 106.95. The stem was plumb and the forefoot carried right forward in a bulbous form. From these parent lines a number of models were constructed with various values of beam draught ratio, etc.

There were two values of beam draught ratio used, namely 2.25 and 3.75.

There were five values of displacement length ratio used, namely 26.60, 53.20, 79.81, 133.02 and 199.52.

There were eight values of longitudinal coefficient used, namely .48, .52, .56, .60, .64, .68, .74 and .80.

Fig. 80 shows relative curves of sectional area used for the eight values of the longitudinal coefficient.

Each of the 80 models was run, its curve of residuary resistance in pounds per ton determined and from the results of the two groups of different beam ratios after cross fairing, Figs. 81 to 120 were plotted.

Each figure refers to a fixed value of $\frac{B}{H}$ and of $\frac{V}{\sqrt{L}}$. It shows contours of residuary resistance in pounds per ton over the range of values of longitudinal coefficient and $\frac{D}{\left(\frac{L}{100}\right)^3}$ most likely to be found in practice. In applying the results of Figs. 81 to 120 for approximate estimates of E.H.P. for beam draught ratios other than 2.25 and 3.75, interpolation of resistance is linear. This is warranted by results of experiments with models from the same parent model and of intermediate beam draught ratio. While not quite exact, it seems sufficiently close to the truth for practical purposes.

4. **Estimates of E.H.P. from Standard Series.** — We are now prepared to calculate curves of E.H.P. for a vessel of any size beam ratio and length within the range covered by Figs. 81 to 120 and from the parent lines of the Standard Series. Table X shows the complete calculations for a vessel of the size, beam ratio and length of the U. S. S. *Yorktown*. For each value of $\dfrac{V}{\sqrt{L}}$ the corresponding figures for the two beam ratios are consulted and columns 2 and 3 filled with the values of $\dfrac{R_r}{D}$ for longitudinal coefficient $= .592$ and $\dfrac{D}{\left(\dfrac{L}{100}\right)^3} = 138.1$. Then in succession columns 5, 4 and 8 are filled as indicated in the headings. Column 6 is filled from Fig. 78.

The correction factor (*b*) for $\dfrac{R_f}{D}$ is obtained as clearly indicated in the heading and column 7 is column 6 × *b*.

The total residuary resistance in pounds per ton is entered in column 9, and column 10 contains the E.H.P. factor by which this must be multiplied to determine at once the E.H.P.

This E.H.P. factor is $.00307\ DV$, but it is convenient to call it $.00307\ D\sqrt{L} \times \dfrac{V}{\sqrt{L}}$. Then (*a*) or $.00307\ D\sqrt{L}$ is calculated and entered in the heading and the values of $\dfrac{V}{\sqrt{L}}$ are found in the first column. Column 11 contains the E.H.P. and column 12 the corresponding values of *V*. Column 10 could be obtained by multiplying column 12 by .00307 D, but the methods indicated in the table will usually be found more convenient in practice.

5. Comparison of Standard Series Estimates with *Yorktown* Model Results. — As illustrating the application of the Standard Series results to estimates of E.H.P. attention is invited to Fig. 121. This shows the E.H.P. curve of the *Yorktown* as determined by experiment with a model of the vessel and the curve of E.H.P. from the Standard Series as calculated in Table X. It is seen that the Standard Series E.H.P. is less than the actual model E.H.P. up to the speed of 18 knots, which is higher than the trial speed of the *Yorktown*. This simply shows that the Standard Series lines are better than those of the *Yorktown*. As a matter of fact, hardly any models of actual ships tried in the Model Basin have shown themselves appreciably superior as regards resistance to the Standard Series and very few have been equal to it. Figs. 76 and 122 show further comparison between actual models and Standard Series results. Fig. 76 shows six E.H.P. curves calculated from six actual models for the *Yorktown* and five variants having the same length and displacement and derived from the *Yorktown* lines but having varying proportions of beam and draught as indicated in the table with Fig. 76.

Fig. 122 shows E.H.P. curves for the same six vessels estimated from the Standard Series results. It is seen that the agreement is reasonably close. The Standard Series generally shows less power than the vessels on *Yorktown* lines, and the curves from it are more closely bunched, but the general features of the two figures are markedly similar.

6. Effect of Longitudinal Coefficient. — Figures 81 to 120, showing the residuary resistance for vessels on the lines of the Standard Series, are worthy of the most careful and attentive study. Attention may be called to one or two of the most obvious features. It is seen that for nearly every speed there is for a given displacement length ratio a distinct minimum of resistance corresponding to a definite longitudinal coefficient. For low and moderate speeds up to $\dfrac{V}{\sqrt{L}} = 1.1$ the best longitudinal coefficient is between .5 and .55. Above this point, however, the optimum longitudinal coefficient rapidly increases, reaching about .65 when $\dfrac{V}{\sqrt{L}} = 1.5$ and being a little greater still when $\dfrac{V}{\sqrt{L}} = 2.00$.

The influence of variation of longitudinal coefficient is greatest below extreme speeds, and it is very great indeed at some speeds. Thus, in Fig. 91, for $\dfrac{B}{H} = 2.25$, $\dfrac{V}{\sqrt{L}} = 1.1$, $\dfrac{D}{\left(\dfrac{L}{100}\right)^3} = 100$, the residuary resistance in pounds per ton for a longitudinal coefficient of .55 is about 6½. But for a longitudinal coefficient of .65 the residuary resistance in pounds per ton is more than doubled — being over 14.

7. Effect of Displacement Length Ratio. — The change in type of the figures with increasing speed length ratio is notable. Thus, for speed length ratio of .75 the contours are nearly vertical in wake of the rather full coefficients which such slow ships would usually have. This means that if we keep length and speed constant and increase displacement, the residuary resistance per ton remains practically constant or the residuary resistance varies as the displacement. Consider now Fig. 100, where the speed length ratio is 2.0. For displacement length ratio = 30 the optimum longitudinal coefficient is about 63 and the residuary resistance in pounds per ton about 51. For the same longitudinal coefficient and a displacement length ratio of 50 the residuary resistance in pounds per ton is about 77. This 77 applies not only to the 20 increase above 30 but to the original 30 as well as that. Though the relative displacements are as 50 to 30, the relative residuary resistances are as 50 × 77 to 30 × 51 or as 3850 to 1530. So an increase of displacement of 66⅔ per cent means an increase in residuary resistance of about 165 per cent.

8. Optimum Midship Section Area. — The displacement, length and longitudinal coefficient being fixed, the area of midship section can be calculated without difficulty. For convenient reference, however, Fig. 123, derived from a series of 2.92 beam draught ratio on the lines of the Standard Series, gives contours of (midship section area) $\div \left(\dfrac{L}{100}\right)^2$ for minimum residuary resistance plotted on speed length ratio and displacement length ratio. From this diagram there may be readily determined in a given case the optimum midship section area as regards residuary resistance. Of course, in practice there are many considerations affecting midship section area besides that of minimum residuary resistance, and the midship section cannot be fixed from considerations of resistance only.

9. **Effect of Length.** — Figs. 81 to 120 do not show directly the effect of variation of length but may be readily utilized to do this.

Thus, suppose it is required to design a vessel of 30,000 tons displacement to be driven at 29 knots. For preliminary work assume $\frac{B}{H} = 3.75$.

Assuming various lengths we use Fig. 78 to determine the corresponding values of the frictional E.H.P. and the Standard Series figures for $\frac{B}{H} = 3.75$ to determine the residuary E.H.P. It is assumed in this preliminary work that it is possible to adopt the optimum cylindrical coefficients.

Fig. 124 shows for the case under consideration separate curves of frictional and residuary E.H.P. and a curve of their sum, or the total E.H.P. all plotted on L. The slow growth of frictional E.H.P. and the rapid falling off of residuary E.H.P. with length are evident. It is seen that the minimum total E.H.P. corresponds to a length of 950 feet. It has already been pointed out that in practice the length should be made less than that for minimum resistance.

Thus, if the vessel were made 850 feet long the increase of E.H.P. would be infinitesimal, and if made 750 feet the increase would be only from 36,500 to 40,200. As the length is made shorter, however, the E.H.P. begins to rise very rapidly. This figure illustrates clearly the enormous effect of length upon residuary resistance. Thus the residuary E.H.P. is a little over 5000 for a length of 950 feet and is 50,000 for a length a little below 600 feet.

It may be noted here that for a case such as that shown in Fig. 124 it would usually be advisable to adopt a longitudinal coefficient above that for minimum resistance. This for several reasons, among which may be mentioned the better behavior in a sea way associated with the fuller ends, and the better maintenance of speed in rough water associated with the smaller midship section.

For a vessel where $\dfrac{V}{\sqrt{L}}$ is large, however, it is usually advisable to make the longitudinal coefficient *less* than that for minimum resistance. Such vessels are nearly all torpedo boats or destroyers, which cruise usually at speeds below their maximum, and it is advisable to save power at cruising speeds by using a longitudinal coefficient a little below that best for maximum speed.

10. Parallel Middle Body Results. — The Standard Series results of Figs. 81 to 120 do not apply to one important type of vessel, namely, the slow vessel of speed length coefficient from .5 to .8 with a parallel middle body. Two questions arise in this connection. First, whether as regards resistance it is advisable to use a parallel middle body, and second, what is the most desirable length for the parallel middle body in a given case?

Experiments were made with models having a midship section coefficient of .96, a ratio of beam to draught of 2.5, various values of displacement length coefficient and three values of longitudinal coefficient, namely, .68, .74 and .80. For each longitudinal coefficient and displacement length coefficient one model was made without parallel middle body and four with parallel middle body. The lengths of parallel middle body expressed as fractions of whole length were as follows:

For .68 longitudinal coefficient, .09, .18, .27, .36.
For .74 longitudinal coefficient, .12, .24, .36, .48.
For .80 longitudinal coefficient, .15, .30, .45, .60.

Curves of residuary resistance were deduced somewhat as in Figs. 81 to 120.

It was found that at low speeds there is a distinct advantage in using parallel middle body. This means, of course, that at these speeds for a given longitudinal coefficient it is advisable to place as much displacement as possible amidships and to fine the ends.

It was found too that when contours of residuary resistance were plotted for a given longitudinal coefficient and speed length coefficient, the abscissæ being percentages of parallel middle body and the ordinates displacement length coefficients, the contours were practically vertical in the vicinity of the optimum length of parallel middle body or that for minimum residuary resistance. In other words, under these conditions the residuary resistance in pounds per ton does not vary much with displacement length coefficient and the latter can be practically eliminated as a variable. Hence, for the purpose in hand the results of the experiments with the models of parallel middle body may be summarized in Figs. 125, 126 and 127 which apply to the three cylindrical coefficients used, namely, .68, .74 and .80. Thus, consider Fig. 126. The abscissæ are values of $\frac{V}{\sqrt{L}}$. One curve shows percentage length of parallel middle body for minimum residuary resistance. The corresponding residuary resistance is given. For convenience, two other curves are given, which show approximately the percentages of parallel middle body greater and less than the optimum, which correspond to residuary resistance ten per cent greater than the minimum. These give an idea of the variations of length of parallel middle body permissible without great increase of residuary resistance.

That the saving by the use of parallel middle body is real is evident from Fig. 128. This gives the three curves of residuary resistance in pounds per ton for the optimum length of parallel middle body from Figs. 125, 126 and 127 and average curves for the same longitudinal coefficients for the Standard Series with no parallel middle body. The lines of the Standard Series appear to be slightly superior to those used for the models with middle body, but even so the saving by the use of the optimum length of parallel middle body is appreciable.

While three coefficients are not enough to fair in exact cross curves on longitudinal coefficient, an approximation can be made from them of ample accuracy for practical purposes, and Fig. 129 shows plotted on speed length coefficient and longitudinal coefficient by full lines contours of optimum length of parallel middle body and by dotted lines corresponding residuary resistance in pounds per ton. It should be understood that the optimum length of parallel middle body shown in Fig. 129 can be materially departed from, as indicated in Figs. 125, 126 and 127, without much increase of residuary resistance.

Particular attention is invited to Fig. 129 which shows how rapidly residuary resistance increases with speed for full models and also how rapidly at speeds above the very lowest it increases with increase of longitudinal coefficient. A judicious selection of a longitudinal coefficient suitable for the speed is just as important for slow vessels as for fast. While hard and fast rules cannot be laid down, experience appears to indicate that few good designers adopt coefficients and proportions for slow ships such that the residuary resistance is much over 30 per cent of the total; and though it is as low as 20 per cent of the total in but few cases, this figure, if it can be attained for low-speed ships, results in vessels which are very economical in service.

15. Squat and Change of Trim

In discussing the disturbance caused in the water by a ship, this question has been touched on, Figs. 45 to 49 showing changes of trim and level for two models at several speeds.

1. **Changes of Level of Bow and Stern.** — It is the practice at the United States Model Basin when towing models for resistance to measure the rise or fall of bow and stern and then plot curves showing the relation between speed and change of level of bow and of stern. These results apply linearly to model and ship at corresponding speeds; that is to say, if the ship dimensions are l times those of the model, the rise of bow of the ship at a given speed will be l times the rise of the model at corresponding speeds.

This fact is taken advantage of in plotting the curves of Figs. 130 to 139, which show for 10 models curves of change of level of bow and stern, the departures of bow and stern from original level being expressed as fractions of length L and plotted not on actual speeds but on values of $\dfrac{V}{\sqrt{L}}$. These curves are then applicable to any size of ship upon the lines of the model from which they were deduced. Actual values of rise and fall can be determined promptly for any

RESISTANCE

speed and length of ship by multiplying by L the values of the curve ordinates for the $\frac{V}{\sqrt{L}}$ values of the ship. Change of trim in degrees can be determined with sufficient approximation by multiplying the difference between the scale values of bow and stern levels by the constant 57.3, the value in degrees of a radian or unity in circular measure. There are given on the face of each figure the values of the displacement length coefficient, the longitudinal coefficient and the midship section coefficient of the corresponding model, thus enabling adequate ideas of its general type to be formed.

The curves of Figs. 130 to 139 show what would happen to vessels that are towed. The propeller suction in the case of screw steamers would cause such vessels when self-propelled to sink more by the stern than indicated, but the difference would not be great.

2. General Conclusion as to Level and Trim Changes with Speed. — The results of Figs. 130 to 139 are typical of results shown by hundreds of other models which warrant the general conclusions below upon the subject of the change of level and trim of vessels under way in deep smooth water.

1. At low and moderate speeds below $\frac{V}{\sqrt{L}} = 1.0$ both bow and stern settle. For short full vessels this bodily settlement is much greater than for long fine vessels.

2. Below $\frac{V}{\sqrt{L}} = 1.0$ about, there is little or no change of trim. In the majority of cases the bow settles a little faster than the stern, particularly for rather full vessels.

3. As speed is increased beyond $\frac{V}{\sqrt{L}} = 1.0$ the bow settles more slowly, reaches an extreme settlement at about $\frac{V}{\sqrt{L}} = 1.15$, and soon begins to rise rapidly, reaching its original level when $\frac{V}{\sqrt{L}} =$ 1.3 to 1.4, and continuing to rise. The stern settles more and more rapidly beyond about $\frac{V}{\sqrt{L}} = 1.2$, and settles much more rapidly

than the bow rises, so that the ship as a whole continues to settle while rapidly changing trim.

4. At about $\frac{V}{\sqrt{L}} = 1.7$ to 1.8 the stern is settling less rapidly than the bow is rising, so that bodily settlement reaches its maximum. The stern does not change its level much beyond $\frac{V}{\sqrt{L}} = 2.0$, while the bow rises always with increase of speed, the result being that the vessel is rising again at speeds beyond $\frac{V}{\sqrt{L}} = 2.0$ about. The center of ordinary vessels will never rise to its original level at any practicable speed; but, since the effect of the passage of the vessel is to depress the immediately surrounding water, it may seem at very high speeds as if the vessel had risen above its original level.

Vessels of special forms and skimming vessels if driven to extreme speeds may rise bodily.

3. **Critical or Squatting Speed.** — The most striking feature of change of level curves is the abrupt change at about $\frac{V}{\sqrt{L}} = 1.2$, the critical speed at which the bow begins to rise and the stern to settle abruptly, causing rapid change of trim.

This "squatting" is often thought to be a cause of excessive resistance. As a matter of fact, it is simply a result of large bow wave resistance. At $\frac{V}{\sqrt{L}} = 1.1$ to 1.2 the first hollow of the bow wave is somewhere near amidships and the second crest somewhere forward of the stern holding it up, as it were. With increase of speed the crest moves aft clear of the stern and the hollow moves aft toward the stern. The stern, of course, drops into this bow wave hollow, causing the "squatting" or rapid change of trim noticed. As speed is increased the hollow in turn moves beyond the stern and the vessel advances on the back of its own bow wave, as it were. The higher the speed, the longer the bow wave and the closer the vessel is to the crest.

It is perfectly true that marked squatting generally means great resistance, because it is the result of an excessive bow wave with a deep first hollow. With no bow wave there would be no squatting, and with slender models having small bow waves squatting is much less marked than for short full models. In every case, however, it is a symptom rather than a cause of resistance.

4. **Perturbation below Critical Speed.** — Figs. 131, 132, 133 and 139 show perturbation in the change of level curves below the critical speed $\frac{V}{\sqrt{L}} = 1.2$. These models are very full ended and have such strong bow waves that as the hollow corresponding to $\frac{V}{\sqrt{L}} = 1.0$ passes the stern it drops into it and the bow rises. Reverse operations take place as the next bow wave crest passes, and then we reach the critical speed, when the stern drops into the bow wave hollow corresponding to $\frac{V}{\sqrt{L}} = 1.2$ and over.

Instead of the pronounced perturbations of quite full models we find for moderately full models the wave hollows and crests passing the stern at speeds below the critical speed cause the curves of change of level to have flat or unfair places. Fig. 135 is a case in point.

For fine models the bow wave is generally so small and the change of level also so small that no effect of the bow wave can be traced in the curves until we reach the critical speed $\frac{V}{\sqrt{L}} = 1.2$.

In considering Figs. 130 to 139 we should bear in mind that the large variations of level and trim shown are for speeds reached by very few vessels.

The curves of Figs. 130 to 139 show changes of level with reference to the natural undisturbed water level, and not with reference to the level of the water in the immediate vicinity of the ship. We have already seen in discussing the disturbance of the water by a ship that, as illustrated in Figs. 45 to 49, the passage of the ship causes disturbances of water level in its vicinity the net result being that on the average there is depression of the water immediately surrounding the vessel.

The changes of level, trim, etc., shown by vessels under way in shallow water differ somewhat from those found in deep water, and will be taken up when considering other shallow-water phenomena.

16. Shallow-Water Effects

1. Changes in Nature of Motion from that in Deep Water. — It is to be expected that as the water shoals the resistance of a ship moving through it will become greater. When the water can move freely past the ship in three dimensions the pressures set up by the ship's motion would naturally be less than when shallowness compels the water to motions approaching the two-dimensional character. Referring to Fig. 21, the greater stream pressures for plane or two-dimensional motion are evident. In shallow water these extra pressures cause waves larger than those in deep water, and in shallow water the lengths of waves accompanying a ship at a given speed are greater than for the same speed in deep water. These are the principal factors differentiating shallow-water resistance from deep-water resistance. There is a third factor, namely, the change in stream velocities past the surface of the ship when in shallow water. This factor would increase resistance somewhat, but its effect would seem to be so small that it is not necessary to consider it since we cannot at present determine with much accuracy the effect of the dominant factor, namely, the change in wave production. We can, however, as a result of experiments with models and full-sized boats get an excellent qualitative idea of the phenomena.

2. Results of Experiments in Varying Depths. — Figs. 140 to 144 show a series of curves of resistance or indicated horse-power. The data from which these curves were constructed came from widely separated sources. The information regarding the German torpedo boat destroyer came originally from a paper by Naval Constructor Paulus in the *Zeitschrift der Vereines Deutsche Ingenieure* of December 10, 1904. Data for the Danish torpedo boats was given by Captain A. Rasmussen, one of the first experimental investigators in this field. The "Makrelen" data was given in *Engineering* of September 7, 1894, and the "Sobjörnen" data in a paper read before the Institution of Naval Architects in 1899. Data for the torpedo boat model was given by Major Giuseppe Rota, R. I. N., in a paper read in 1900 before the Institution of Naval Architects, the experiments with the model having been made in the Experimental Model Basin at Spezia, Italy. Information from which the curves for the Yarrow destroyer were deduced was given in a paper before the Institution of Naval Architects in 1905 by Harold Yarrow, Esq. In Mr. Yarrow's paper curves of E.H.P. were given as deduced from model experiments in the North German Lloyd experimental basin at Bremerhaven.

Each curve refers to a definite depth of water, which has been expressed as a fraction of the length of the vessel. Furthermore, speed has been denoted not absolutely but by values of $\dfrac{V}{\sqrt{L}}$.

3. Deductions from Experimental Results. — Examining the curves, which range from those for a 145-pound model to those for a 600-ton destroyer, and bearing in mind the varying depths expressed as fractions of the length, we seem warranted in concluding that in a depth which is a given fraction of the length the perturbations occur at substantially the same values of $\dfrac{V}{\sqrt{L}}$ regardless of the absolute size. The reason for this must be sought in the relation between the length of a wave traveling at a given speed in a given depth of water and length of vessel.

By the trochoidal theory the formula giving wave speed in shallow water is

$$v^2 = \frac{\epsilon^{4\pi \frac{d_0}{l}} - 1}{\epsilon^{4\pi \frac{d_0}{l}} + 1} \cdot \frac{gl}{2\pi},$$

where l is length of wave in feet, d_0 is depth of water in feet and v is speed of wave in feet per second.

Now let L denote length of ship in feet and put $l = cL$.

Also let V denote common speed of ship and wave in knots.

Then $V = v\,\dfrac{3600}{6080}$. Substituting, reducing and putting $g = 32.16$ we have

$$\frac{V}{\sqrt{L}} = 1.34\,\sqrt{c}\,\sqrt{\frac{\epsilon^{\frac{4\pi}{c}\frac{d_0}{L}} - 1}{\epsilon^{\frac{4\pi}{c}\frac{d_0}{L}} + 1}}.$$

Fig. 145 shows contour curves of equal values of c plotted on axes of $\dfrac{d_0}{L}$ and $\dfrac{V}{\sqrt{L}}$. Fig. 145 also shows in dotted lines curves deduced somewhat arbitrarily from Figs. 140 to 144 and other data showing the loci of the points at which increase of resistance due to shoal water becomes noticeable, attains its maximum and dies away.

The data is not thoroughly concordant, and the dotted curves of Fig. 145 should be regarded as a tentative attempt to locate regions rather than points. The broad phenomena, however, are clear. A high-speed vessel in water of depth less than her length will at a given speed in a given depth begin to experience appreciably increased resistance as compared with its resistance in deep water. The increase of resistance above the normal becomes greater and greater as speed increases until it reaches a maximum. This maximum appears to be at about a speed such that a trochoidal wave traveling at this speed in water of the same depth is about $1\frac{1}{4}$ times as long as the vessel. As the vessel is pushed to a higher speed the resistance begins to approach the normal again, reaches and crosses the normal at about the speed indicated in Fig. 145, and for higher speeds the resistance in shallow water is less than in deep water.

It was at one time supposed that the speed for maximum increase in resistance was that of the wave of translation. This, however, as illustrated in Fig. 145, holds only for water whose depth is less than $.2\,L$. For greater depths the speed of the wave of translation rapidly becomes greater than the speed of maximum increase of resistance.

There are obvious advantages in the model-basin method of investigating this subject. Consider, for instance, Fig. 144 showing actual falling off of resistance beyond the critical speed in the curves for the Yarrow destroyer which were obtained by model-basin experiment. This remarkable feature would never be detected on a full-scale trial of an actual destroyer, because if such a vessel were forced to surmount the hump it would leap the gap, as it were, and show a sudden jump in speed. Theoretically if the depth of water were absolutely uniform it would be possible after the jump in speed to gradually throttle down until the boat would be working in the hollow, but the chance of this ever being done, unless it were known that the hollow should be there, is infinitesimal.

4. Shallow-Water Experiments at United States Model Basin. — That the hollow really exists, as shown in the curves for the Yarrow destroyer, is confirmed by published results of other model-basin shallow-water experiments and by a number of carefully made experiments in the United States Model Basin.

Fig. 147 shows curves of resistance and change of trim of the model of a fast scout in various depths of water. The model was 20 feet long on L.W.L., with $2'.268$ beam and $0'.842$ mean draught. It displaced in fresh water 996 pounds. The corresponding speed of the model for 30 knots speed of the full-sized ship would be only 6.61 knots, but the experiments were carried to a much higher speed as a matter of interest.

The sudden and peculiar drops in the shallow-water curves are very marked. It is seen that they are accompanied by peculiar corresponding perturbations in the curves showing change of trim or change of level of bow and stern. We have from Fig. 147:

	18″	24″	36″
Depth of water....................................			
Speed of maximum % increase of resistance, knots...	4.02	4.54	5.21
Trochoidal wave lengths — above speed and depth...	$25.5'$	$23.5'$	$21.4'$
Speed of hollow in resistance curve, knots...........	4.60	5.05	5.95
Speed of wave of translation or trochoidal wave of infinite length in the depth of water, knots........	4.11	4.75	5.82

The general features of Fig. 147 agree closely with results of trials of other models in shallow water at the United States Model Basin. Some peculiar wave phenomena appear in such trials. In running such models in deep water or in shallow water at speeds well below that of the hump the disturbance set up in the water is inappreciable a short distance ahead of it. But at about the speed of the hump the wave at the bow tends to manifest itself as a crest extending straight across the basin and well ahead of the bow — as much as 8 or 10 feet. As the speed is increased this singular manifestation disappears, and again there is no appreciable disturbance ahead of the model. These phenomena have not been given careful investigation. A reasonable explanation of the sudden drop of the resistance curve would be that it corresponds to the wave of translation, which advances with less demand upon the model for energy to maintain it than was the case at a slightly lower speed when the wave system was being built up even ahead of the model.

At the higher speeds the waves are forced waves, necessarily departing widely from trochoidal waves. It should be remarked that the high "deep water" resistance of the model at speeds in the vicinity of 8 knots may be in part due to the limited depth (14 feet) of the basin, but is probably mostly due to the appearance of the last normal deep-water hump of resistance curves. The hump which appears below 6 knots in 46 inches depth is found at about 8 knots in 14 feet depth.

5. Shallow-Water Resistance for Moderate and Slow Speed Vessels. — The case of greatest practical interest is that of the vessel of moderate speed — say capable of a deep-water speed in knots of $.9 \sqrt{L}$ or less. Such a vessel in shallow water cannot be pushed beyond the last hump of her resistance curve, and hence always loses speed in shallow water. For such vessels we would like to know the least depth of water in which resistance is not appreciably increased or speed appreciably retarded and the amount of increase of resistance in water that is shallower.

Results of experiments bearing directly on the first question were published in 1900 in a paper before the Institution of Naval Architects by Major Giuseppe Rota. Major Rota experimented with models of five vessels, one being the torpedo boat model, whose results are given in Fig. 143. Each model was run in various depths of water and the results carefully analyzed for the purpose of determining the depth at which increased resistance began.

For the purpose of analysis and deducing results applicable to other vessels it is important to determine in connection with such experimental results the fundamental variables, as it were. For instance, in this case shall we connect the depth of water with the length, the beam or the draught of the ship? We have seen that for high-powered vessels we were led to the use of the ratio between depth of water and length of vessel, which gives satisfactory results as regards determination of critical points, etc. Consideration, however, appears to indicate that for the vessel of moderate speed it would probably be better to use the ratio between depth of water and mean draught of ship, allowing the length factor to come in through the speed-length coefficient.

While Rota's models could, of course, each be expanded to represent any number of ships, he gives one size of ship for each as shown in the table below.

Model No.	1	2	3	4	5
Displacement of ship in tons..	12,000	8000	6000	3000	1000
Length of ship in feet........	408	385	361	380	263
Beam of ship in feet.........	75.5	67	55	40.3	28
Mean draught of ship in feet .	26.6	21.4	20.2	13.8	9.6
Block coefficient.............	.51	.50	.50	.49	.43

Taking Rota's curves giving the depths for no increase of resistance for various speeds of the above ships and replotting them to express in each case a relation between $\frac{V}{\sqrt{L}}$ and the depth of water expressed in draughts of the ship, we have the results shown in Fig. 146. It is seen that for each model the locus thus plotted is reasonably close to a straight line and that the dotted line is reasonably close to the average of the five up to the speeds not greater than $\frac{V}{\sqrt{L}} = .9$. Curiously enough, the two finer models fall above the dotted line. This, however, is probably due to the fact that they are vessels of distinctly shallow-draught type, and because of that, in spite of their fineness, need a depth of more draughts than vessels of deeper-draught type. A scrutiny of Rota's results, however, indicates that for models 4 and 5 the decrease of depth from that of lines 4 and 5 in Fig. 146 to that of the dotted line will involve in practice an increase of resistance barely perceptible. Then Rota's experiments may be fairly summarized by the straight line of Fig. 146. If H denotes the draught, it is seen from the diagram that this line gives us the relative minimum depth for no increase of resistance $= 10H \frac{V}{\sqrt{L}}$. This formula giving minimum depth for no increase of resistance applies, strictly speaking, only to Rota's five models, but it is seen that they cover the range of usual proportions for models of a fine block coefficient.

The formula, however, has been found to apply satisfactorily to models of block coefficient higher than .5 tested in the United States Model Basin. One model of block coefficient slightly above .65 was tried in various depths and the formula found to apply satisfactorily.

To sum up, I think that the above formula from Rota's experiments may be confidently applied:

1. To vessels not of abnormal form or proportions up to a block coefficient of .65.

2. For speeds for which $\dfrac{V}{\sqrt{L}}$ is not greater than .9.

The formula may be of use beyond the limits indicated above, but in such cases needs to be applied with caution and discretion.

6. Trial Course Depths. — As illustrative of the little importance attached to this question until a comparatively recent date, Major Rota in his 1900 paper states: "Stokes Bay, where British ships used to undergo their speed trials, is only 59 feet deep; the official measured mile at the Gulf of Spezia, Italy, is about 62 feet deep; the measured miles at Cherbourg and Brest are 49 and 59 feet respectively." Such depths are now regarded as entirely inadequate and no speed trials of large ships are regarded as accurate unless made in deep water. Curiously enough, however, as indicated in Fig. 145, the shallow course exaggerates the speed of the very fast vessel, and there are many torpedo craft in existence whose full-speed trials were held on shallow courses with resulting speeds greater than would have been attained in deep water.

7. Percentage Variations of Resistance in Shallow Water. — Coming now to the question of the actual increase of resistance of a given vessel in water of a given depth, it is necessary again to make a distinction between the vessel of very high power and speed and the vessel of moderate speed. For the former it is probably best, as before, to use as the governing variable the ratio between depth and length, $\dfrac{d_0}{L}$. For the latter it still seems best to use the ratio between depth and draught, $\dfrac{d_0}{H}$. For either type, expressing the speed by $\dfrac{V}{\sqrt{L}}$, we are able for each vessel or model for which there is adequate experimental information to draw contours on $\dfrac{V}{\sqrt{L}}$ and ratio between depth and length or depth and draught as the case may be, which show percentages of increase over deep-water results. For the very high-speed vessels percentages of decrease will also appear. This work at best can be only a tolerably good approximation, and hence we assume in it that the law of comparison applies fully to the total model resistance. Figs. 148 to 153 are percentage increase diagrams, the type of vessel being indicated in each case.

The diagrams for the high-speed vessels show percentages of decrease. For the moderate-speed vessels the percentage increase of resistance goes up rapidly with increase of displacement length coefficient. While Figs. 151, 152 and 153 cannot be said to cover the ground as would be desirable, they will be better than nothing and of help in many cases.

Inland navigation is mostly smooth-water, shallow-water navigation, and there is great need of a complete investigation into the features of form affecting shallow-water resistance. While we know quite well the general features of the form best adapted to speed in deep water in a given case we do not know the same thing for shallow water. It appears probable, however, that if we wish to make 12 knots in shallow water and are considering various models, that one which will drive easiest in deep water at a higher speed — say 15 knots or so — will drive easiest in shallow water at the 12-knot speed. If high speed is to be attempted in inland navigation there are practical advantages in length which would be excessive for deep-water work. Wave making, with the resulting wash at banks and piers, should be kept as low as possible for boats in river service.

8. Shallow-Water Influence upon Trim and Settlement. — Fig. 147 shows the curves of the settlement of bow and stern of a scout model in shallow water. It is seen that the shallower the water the lower the speed at which marked change of trim begins, and within the limits of practicable speed the greater the change of trim.

For speeds above those at all possible the trim changes would not very greatly depart from those for deep water. We are more concerned in practice, however, with settlement and change of trim at low speeds, corresponding to those at which shallow channels would be traversed. Fig. 147 shows that at such speeds the effect of shoal water is simply to increase the settlement of both bow and stern. In its broad features, Fig. 147 is fairly typical of change of trim results in shoal water for a number of other models. We may say that the effect of shoal water upon a vessel under way is to increase the natural settlement of both bow and stern at low speed. The shallower the water the lower the critical speed at which squatting or excessive change of trim begins and the greater the change of trim. At high speeds the shallower the water the more the stern settles and the more the bow rises. At extreme speeds, however, the stern does not appear to settle or the bow to rise so far as in deep water. It is interesting to note in Fig. 147 the peculiar perturbations in the change of level curves and the evident close connection between them and the remarkable drops in the resistance curves.

9. **Increase of Draught in Shallow Channels.** — In practice there are very few vessels of sufficient power to attain high speed in shallow water, and those that have the power would very seldom use it in shallow water, so that the behavior of vessels as regards settlement under way at moderate speed in shallow channels is of more practical importance than their possible behavior at excessive speeds.

A very interesting investigation of this question was made in connection with the channel of New York Harbor, and was described in detail by Mr. Henry N. Babcock in *Engineering News* for August 4, 1904. This channel was constantly used by large steamers passing in and out with very little to spare between their keels and the bottom of the channel. There were repeated complaints from such vessels that they had touched bottom in places where the officers in charge of the channels were unable to discover spots shoaler than the still-water draught of the steamers. The observations were confined to large transatlantic steamships passing out of New York, averaging over 550 feet in length. They were made at three points, one where the channel was 80 to 100 feet deep, one where the low-water channel depth was from 31.1 to 32.5 feet, and a third where the low-water depth was from 31 to 34.5 feet.

The general scheme of most of the observations was to determine the height above water of marks on the bow and stern before the steamer left her pier. Then as the steamer passed the observing station the level of these marks was determined with reference to the station, and as soon as possible after the passage of the vessel the water level was determined with reference to the observing station. Considering all the circumstances, exact observations are obviously not possible, but after making ample allowance for possible errors of observation Mr. Babcock's report demonstrates conclusively that vessels of the type considered when under way in channels settle both at bow and stern, and the shoaler the water and higher the speed the more they settle. It was not practicable from the results to formulate fully conclusions connecting amount of settlement with size and type of vessel, speed and depth of water, but Mr. Babcock, upon analyzing the results, concluded that for vessels of the large transatlantic steamship type the increase of draught in feet, when still water clearance under their keels was less than about 10 per cent of the draught, would be $\frac{1}{5}$ the speed of the ship in miles per hour. For a natural clearance of some 30 per cent of the draught the increase in feet would be about $\frac{1}{10}$ the speed of the ship in miles per hour, and for intermediate clearances intermediate fractions should be used.

Further observations of the character reported by Mr. Babcock on the settlement of vessels under way, not only in shallow channels but in canals, would be of much interest and practical value.

17. Rough-Water Effects

1. **Causes of Speed Reduction.**—The effect of rough water upon speed is like the effect of foulness of bottom — almost impossible to reduce to quantitative rules. The very real and material reduction of speed of vessels in rough weather is of universal experience.

This, however, is not always due to increased resistance alone. The motion of the ship may render it impossible to develop full power. The danger of racing may render it inadvisable to use full power. The disturbance of the water reduces the efficiency of the propellers. The conditions may render it impossible to use full speed without risk of dangerous seas coming on board.

2. Features Minimizing Speed Reduction. — The increase of resistance in rough water is under practical conditions largely a question of absolute size. Waves 150 feet long and 10 feet high would not seriously slow a 40,000-ton vessel 800 feet long.

A vessel of a few hundred tons 120 feet long would find them a very serious obstacle to speed. Pitching enters into the question of rough-water speed as a very important factor.

When conditions are such as to produce severe pitching, speed goes down very rapidly. Pitching exaggerates nearly all causes of speed loss. Not only is actual resistance rapidly increased but racing is caused, the propeller loses efficiency and more water comes on board.

If it were possible to devise a vessel which would not pitch it would lose much less speed in rough water than one that does pitch; but though many naval architects have strong opinions on the subject there is no agreement among them as to the features of model which minimize pitching. The preponderance of opinion is probably in favor of the U-bow type and rather full bow water lines. But pitching is unfortunately largely a question of conditions. Under certain conditions of sea, course, and speed one type may be superior and under slightly changed conditions distinctly inferior.

Apart from absolute size there appears, however, to be one broad consideration which is of some value as a guide. Suppose we have two 20-knot vessels, A and B, of about the same power and such that at 22 knots A offers distinctly less resistance than B. There is little doubt that on the average A would lose less speed in rough water than B.

When for a vessel intended for a certain service it is necessary to allow in the design for the effect of rough water upon speed there is only one safe method to follow — namely, to allow a reduction from smooth-water trial conditions to rough-water service conditions based upon actual experience with previous vessels in the service.

18. Appendage Resistance

1. Appendages Fitted. — Substantially all that has been said about resistance hitherto refers to the resistance of the main body or hull proper. There are found on actual ships appendages of various kinds, such as rudders, bar keels, bilge keels, docking keels, shaft swells, shafts, shaft struts, propeller hubs and spectacle frames, or shaft brackets or bosses. Shaft tubes, or removable tubes around the outboard shafts, are seldom fitted nowadays.

The appendages fitted vary. Thus, a single-screw merchant ship with flat keel will have practically no appendage except the rudder, the slight swell around the shaft having hardly any effect. For such a vessel the appendage resistance would seldom be as much as 4 or 5 per cent of the bare hull resistance.

A twin screw vessel with large bilge and docking keels and perhaps two pairs of struts on each side may have an appendage resistance as much as 20 per cent of the bare hull resistance.

Appendage resistance is largely eddy resistance and can be kept down to the minimum only by very careful attention to details and the application of adequate fair waters wherever needed.

2. Resistance of Bilge and Docking Keels. — Bilge keels and docking keels should follow lines of flow and be sharpened at each end. When this is done it is generally found in experiments upon models that the additional resistance due to them is not greater than that due to the additional surface alone. In fact the additional resistance is sometimes found to be less than that due to the additional wetted surface. Mr. Froude found a similar result in his full-sized *Greyhound* experiments. While if bilge keels and docking keels are properly located and fashioned the additional resistance may be taken as that due to their wetted surface only, the wetted surface they add is often very considerable.

In models bilge keels may be located at appreciable angles with the natural lines of flow without greatly augmenting resistance beyond that due to their surface, but it does not follow that the same result would be found in the full-sized ships. It is necessary to be cautious in applying the Law of Comparison to eddy resistance. There is little doubt that the law applies to the

Eddy Resistance behind a square stern post, for instance. Here the eddying for model and ship is found in each case over corresponding areas.

But in the case of a bilge keel located across the lines of flow we may readily conceive that there may be but little eddying around the model bilge keel and a great deal around the full-sized bilge keel. This because the pressure of the atmosphere remaining constant the total pressure around the full-sized bilge keel is not increased in the proportion required to insure compliance with the Law of Comparison.

3. **Resistance of Struts.**—Probably struts and spectacle frames are the appendages to which the most careful attention must be paid from the point of view of resistance. Experiments with a number of strut arms of elliptical section appear to indicate that the resistance in pounds per foot length may be expressed with fair approximation for areas from 40 square inches to 175 square inches by the following semi-empirical formula:

$$R = \frac{C}{1000}(A + 40)V^2.$$

Where R is resistance in pounds per foot length, V is speed through the water in knots and A is area of cross section of strut in square inches. The coefficient C depends upon the ratio between B, the thickness of the strut section, and L, its width in direction of motion. The table below gives values of C for various values of $\frac{L}{B}$.

$\frac{L}{B}$	3	4	5	6	7	8	9	10	11	12
C	1.880	1.318	1.073	.940	.858	.801	.762	.736	.720	.714

From the point of view of resistance only, the best ratio of breadth to thickness would be 10 or over, but as the wide, thin strut requires more area for a given strength, it follows that the best all-round ratio would be somewhat smaller, say from 7 to 9.

Even this ratio is not very often reached in practice, the tendency apparently being to make strut arms much narrower and thicker than they should be.

As regards shape of section, model experiments indicate that a pear-shaped section, or a section of rounding forward part and sharp after part, offers the least resistance. Such a section may show model resistance as much as 10 per cent below the elliptical section.

There is doubt, however, whether this holds for full-sized struts for high-speed vessels. Study of Fig. 16 would seem to indicate that at sufficiently high speeds there must be eddying over all the rear half of any strut, in which case the thickness of the strut should be reduced to a minimum. From this point of view, if a strut of given width and area is to have the minimum thickness for a given type of head the rear portion should be made of parallel thickness and cut off square. Furthermore, from this point of view, if air were piped to the rear of a strut the resistance would be decreased. This question of strut resistance is worthy of further careful experimental investigation. Pending this, the approximate formula and coefficients above for elliptical struts may be used, and it may be assumed that the elliptical form is about as good as any. For moderate speeds the rear portion of the strut may be brought to a sharp edge, but for high speeds this refinement will probably be of little use.

4. Resistances of Propeller Hubs. — Behind the strut hub the propeller hub is fitted, and for propellers with detachable blades is usually larger than the strut hub. About all that can be done for the propeller hub is to fit a conical fair-water behind it. Model experiments show that a long fair-water, say of length about twice the diameter of the propeller hub, offers materially less resistance than a short fair-water of length say about one-half the diameter of the propeller hub.

While there is some doubt whether the long fair-water would show up so well in comparison on the full-sized ship, the length of fair-water should not be skimped.

With quick running propellers the objections to large hubs have become more evident and there is a tendency to use solid propellers with small hubs. From the point of view of appendage resistance, these are distinctly preferable to large hubs.

5. Resistance of Spectacle Frames or Propeller Bossing.—In merchant practice, struts are not much used for side screws, being replaced by spectacle frames or propeller bossing.

These appendages, if well formed, offer less resistance than thick struts with the bare shafts, etc., but in many cases wide, reasonably thin struts would offer less resistance than shaft bosses. Shaft bosses are, however, usually regarded as giving better security to the shaft, and certainly give access to a greater portion of its length. They absorb much more weight than struts. The angle of the web of a shaft boss may vary a good deal from what may be called the neutral position, or position where it is edgewise to the flow over the hull without very great effect upon the model resistance, but there is a little doubt that the full-sized ship will be prejudicially affected if the shaft boss webs depart too far from the neutral position. Eddying is liable to appear in the case of the full-sized ship which does not occur in the case of the model.

The angle of such webs has a powerful influence upon the stream line motion in the vicinity of the stern. A vertical web or a horizontal web tends seriously to obstruct the natural water flow and drag more or less dead water behind the ship. It seems to be usually the tendency from structural considerations to work the shaft boss webs somewhere near the horizontal. From the point of view of resistance alone a 45° angle for the rear edge may not be too great. This is another case where conflicting considerations necessitate a compromise. The determination of after lines of flow over the hull will greatly facilitate the determination of the most suitable shaft boss arrangements.

6. Allowance for Appendages in Powering Ships.— In estimating from model experiments the effective horse-power of a ship with appendages the methods are the same as for the bare hull. From the total model resistance the frictional resistance for the total wetted surface including appendages is deducted and the remaining or residuary resistance treated by the Laws of Comparison. From what has been said in discussing appendage resistance, it is evident that estimates of E.H.P. with appendages are apt to be less accurate than estimates of the net or bare hull E.H.P. unless care has been taken so to shape appendages that they do not develop in the full-sized ship eddies which have no corresponding eddies in the case of the model.

In practice, it is customary and almost necessary to power a new design from model experiments with bare hull only. This is readily done by using for the ratio between the bare hull E.H.P. and the I.H.P. of the ship with appendages a conservative coefficient of propulsion based upon coefficients of propulsion actually obtained from past experience with vessels reasonably similar as regards appendages to the case under consideration.

CHAPTER III

PROPULSION

19. Nomenclature Geometry and Delineation of Propellers

1. Definitions and Nomenclature. — A screw propeller has two or more blades attached at their inner portions or roots to a hub or boss, which in turn is secured upon a shaft driven by the propelling machinery of the ship. Figs. 154 to 157 show plans of a three-bladed propeller for a naval vessel. This is a true screw — that is, the face or driving face is a portion of a helicoidal surface of uniform pitch. A helicoidal surface of uniform pitch is the surface generated by a line — the generatrix — at an angle with an axis which revolves about the axis at a uniform angular rate and also advances parallel to the axis at a uniform rate. A cylindrical surface concentric with the axis will cut such a helicoidal surface in a helix. The pitch of the helicoidal surface is the distance which the generatrix moves parallel to the axis during one complete revolution. Figs. 154 to 157 show a three-bladed right-handed propeller — that is, a propeller which, viewed from aft, revolves with the hands of a watch when driving the ship ahead. The various portions of a propeller are indicated in the figures, such as the face and back of the blades, the leading edge and the following edge, the tip and the root. Since in practice the back of each blade is its forward surface, care must be taken to avoid confusion.

This result will be obtained by avoiding such expressions as "forward face," "after face," etc., and adhering to the terms "face" and "back." The word "face" will always denote the driving face or the face which pushes the water astern when the propeller is in action, while the word "back" naturally denotes the surface opposite the face.

While a true screw as already indicated is a screw propeller whose blade faces are all portions of helicoidal surfaces of the same pitch, there are many variants from the true screw.

Each point of the face may have its own pitch, which may be defined as the distance parallel to the shaft axis which an elementary area around the point would move during one revolution around the shaft if it were connected to the shaft by a rigid radius and working in a solid fixed nut. Fig. 158 shows two views of a small elementary area LL connected to the shaft axis O by a radius r. This area makes an angle with the perpendicular to the axis called the pitch angle and denoted by θ in Fig. 158. If p denote the pitch of LL, during one revolution in a solid nut its center would advance along the helix $OCCD$, to the point D at a distance p along the axis from O. If then we unroll the cylinder of radius r, upon which has been traced the helix $OCCD$, this helix will become the straight line OP of Fig. 158, while $PM = p$, the pitch.

$$OM = 2\pi r \text{ and } \tan\theta = \frac{p}{2\pi r}.$$

There are several typical variations of pitch which are used more or less for actual propellers. Thus if the pitch increases as we pass from the leading to the following edge, the blade is said to have axially increasing pitch. If the pitch increases as we go outward, the blade is said to have radially increasing pitch. If the pitch decreases as we go outward, the blade has radially decreasing pitch. A blade may have pitch varying both axially and radially.

Pitch of the blade face only has been considered in the above, and in an ideal blade of no thickness that is all that need be considered; but for actual blades we need to consider the pitch of the back of the blade as well. Evidently each point of the back of an actual blade has a distinctive pitch. For blades such as shown in Figs. 154 to 157, where the face has uniform pitch and the blade sections are of the usual ogival type, the pitch of the center of the blade back is the same as the pitch of the face. The pitch of the leading portion of the back is less; and of the following portion greater than the face pitch. These pitch variations over the blade back have important effects upon propeller action.

The ratio between pitch and diameter is called pitch ratio, and the ratio between diameter and pitch is called diameter ratio. Each point of a blade has, of course, its own pitch ratio and diameter ratio, but these expressions are also used in reference to the propeller as a whole. When so used the diameter referred to is the diameter of the screw or of the tip circle, and the pitch is the uniform pitch of the face for a true screw and an assumed average face pitch for a screw of varying pitch.

There are two other ratios which it is convenient to define here. Fig. 159 shows a radial section through the center of a blade of very common type by a plane through the axis. This plane intersects back and face of the blade in two straight lines, which, prolonged through the hub to the axis, cut it at C and A respectively.

The ratio $\dfrac{CA}{\text{Diameter}}$ is called the blade thickness ratio and is evidently constant for similar propellers, whatever their size.

The blade section in Fig. 159 is shown raking aft, the total rake reckoned along the mid-thickness of blade sections being in the figure BO. Then $\dfrac{BO}{\text{Diameter}}$ is called the rake ratio. It is reckoned positive for after rake and negative for forward rake.

Propellers do not in practice move through the water as through a solid nut. They advance a distance less than their pitch for each revolution. Under given conditions of operation the distance advanced is the same for each revolution, hence the path of each element is a helix and can be developed into a straight line. Recurring to Fig. 158, $= OC_1C_1D_1$ is the helical path of LL with slip and OS the development of this helix. As before, POM is the pitch angle θ. The angle POS is called the slip angle and will be denoted by ϕ. Fig. 158 may also be regarded as a diagram of velocities, OM being the transverse or rotary velocity of the element and MS its velocity parallel to the axis. MS is often called the speed of advance, and MP, or the speed for no slip, is called the speed of the propeller, being the pitch multiplied by the revolutions. Then PS is the speed of slip or the slip velocity. Slip is usually characterized, however, by the ratio $\dfrac{PS}{PM}$, or the ratio between the speed of slip and the speed of the propeller. This is properly called the slip ratio, or slip fraction. It is also commonly and conveniently called simply the slip and expressed as a percentage instead of a decimal fraction. Thus when we say, for example, that a propeller works with a slip of 15 per cent we mean that

$$\frac{\text{Speed of Propeller} - \text{Speed of Advance}}{\text{Speed of Propeller}} = .15.$$

Sometimes we need the ratio

$$\frac{\text{Speed of Advance}}{\text{Speed of Propeller}},$$

and this may conveniently be designated the speed ratio.

2. Delineation. — In practice a propeller is usually delineated as in Figs. 154 to 157, by projections of the blades in at least two directions, — an expansion of a blade and sections of a blade. Views and sections are also shown as necessary to determine the hub of propeller with solid hubs and the hub and blade flanges and bolting of propellers with detachable blades.

It will be observed that the faces of the sections in Fig. 155 all radiate from a fixed point on the axis, called the pitch point. This is a more or less convenient arrangement. Referring to Fig. 160, suppose p is the pitch of a blade at the radius $OA = r$. Lay off $OP = \frac{p}{2\pi}$. Then $\tan OAP = \frac{p}{2\pi} \div r = \frac{p}{2\pi r}$. But from Fig. 158 $\frac{p}{2\pi r} = \tan \theta$ where θ is the pitch angle or the angle which the element makes with a transverse plane. Hence in Fig. 160 OAP and the corresponding angles at the other radii are the pitch angles at the radii in question.

Figs. 154 to 157 refer to an ordinary true screw of oval blade contour with a rake so small that it is practically negligible. Much more complicated forms are used sometimes, the complications involving varying pitch, curved radial sections, extreme rake forward or aft, lopsided or unsymmetrical blade contours, and various types of blade sections. Some forms of propellers are difficult problems in descriptive geometry. There does not seem to be any benefit in practice from complicated forms of propellers and no attempt will be made to take up the problems of their delineation.

3. **Area and its Determination.** — The question of propeller area is a very important one. There are various areas considered in connection with a propeller. When we speak of the blade area of a propeller we generally mean what is called the helicoidal area, or the actual area of the helicoidal faces of the blades. As it happens, however, a helicoidal surface cannot be developed into a plane so the helicoidal area of a propeller cannot be determined exactly. The area we determine is what is called the developed area, the blade face being developed into a plane by a more or less approximate method.

The disc area of a propeller is the area of the circular section of its disc or the area of the circle touching the blade tips.

The projected area is the area of the projections of the blade faces upon a transverse plane perpendicular to the axis.

The ratio between the developed and disc areas of a propeller is sometimes called the disc area ratio.

The ratio Projected Area ÷ Disc Area is also frequently used and is of more practical value than the ratio Developed Area ÷ Disc Area.

While the helicoidal face of a propeller blade cannot be developed exactly into one plane it can be so developed with such slight distortion that the resulting surface is an approximation amply close for practical purposes.

Suppose we cut the helicoidal surface of a blade face by a cylinder concentric with the axis. It will cut a helix from the helicoidal surface. If now we pass a plane tangent to the helicoidal surface at its center, it will cut the cylinder in an elliptical arc. If then we take that portion of this elliptical arc whose rearward projection is the same as that of the actual helix of the blade face we will have an arc of very nearly the same length as the helix. Then if we take a series of such arcs, swing them into a common plane and join their extremities by a bounding curve, we shall have a developed surface which is very close to the actual helicoidal surface in area.

Fig. 160 shows the construction, O is the center, P the pitch point, OA the radius of a cylinder. Let BBB be the projected blade. Then the cylinder of radius OA cuts BB at C. The plane at A tangent to the helicoidal surface makes with the axis the angle OPA — the complement of the pitch angle. The minor semiaxis of the ellipse which it cuts from the cylinder is OA. The major semiaxis is $= \dfrac{OA}{\sin OPA} = AP$. Draw the elliptical arc AD with major axis of length AP and minor axis OA in length and position. Then draw the horizontal line CD meeting the elliptical arc at D. D is a point on the developed blade, and by determining a series of such points and drawing a line through them we obtain the developed contour $EDBEE$. Suppose now we draw AF horizontal through A and make AF equal in length to the elliptical arc AD. A line through a series of points such as F will give what may be called the expanded contour. It is denoted in the figure by $HFBHH$. The developed area is usually taken as $BEEKEDB$. The expanded area, $BHHKHFB$, is very close to the developed area.

The developed area obtained by the above method is slightly smaller than the true area. The elliptical arcs are not very easy to draw in practice and a simple method is to use arcs of circles with radii which are the radii of curvature of the ellipses. Thus draw PM at right angles to AP and cutting AO produced at M. Then M is the center of curvature of the ellipse at A, and instead of drawing the ellipse we may draw a circular arc of radius MA. The developed area thus determined is slightly greater than the exact helicoidal area, the area using the exact ellipses being slightly less. But the area determined using the circular arcs is a closer approximation to the true area, particularly for broad blades.

In practice we generally assume the developed contour, making it any desired shape, deduce the projected contour by reversing the method of development described above, and from the projected contour deduce by the methods of descriptive geometry the other projections desired. A very common and very good contour for the developed blade is an ellipse touching the axis, having the radius as major axis and the expanded breath of blade at mid-radius as minor axis. In the vicinity of the hub the ellipse is departed from as necessary to make a good connection.

4. Coefficients of Area for Elliptical Blade. — Fig. 161 shows an elliptical developed blade contour with major axis equal to the propeller radius. The radius of hub is $\frac{2}{10}$ that of the blade. There is shown dotted a rectangular area touching the hub and tip circle and of width such that its area is the same as that of the elliptical blade outside the hub. Then the width of this rectangle is called the mean width of the blade.

It is convenient usually to use the diameter as the primary variable when dealing with propellers, so we naturally express the mean width as a fraction of the diameter.

The ratio (mean width of blade) ÷ (diameter of propeller) is called the mean width ratio and is denoted by h.

This mean width ratio characterizes a blade very definitely and it is convenient to express many other features by its use. For the elliptical blade with hub diameter $\frac{2}{10}$ of the propeller diameter let l denote the maximum width or minor axis of the ellipse. Then we have mean width ratio $= h = .842 \dfrac{l}{d}$, or $l = 1.188\ hd$.

If n denote the number of blades we have the total blade area or Developed Area $= .4\ nd^2h$.

The projected area for a given developed area depends upon the pitch ratio, which denote by a. For values of a found in practice, say from $a = .6$ to $a = 2.0$, the projected area for the elliptical-bladed propeller of hub diameter .2 of the propeller diameter is given with close approximation by the formula,

Projected Area $= (0.4267 - 0.0916a)\ nd^2h$.

From the above we have the following additional ratios for values of a between .6 and 2.0:

Projected Area ÷ Developed Area $= 1.067 - .229a$.
Developed Area ÷ Disc Area $= .509\ nh$.
Projected Area ÷ Disc Area $= (.543 - .1166a)\ nh$.

Fig. 162 shows contours of the ratio (Projected Area) ÷ (Disc Area) for elliptical three-bladed propellers.

While the above formulæ and Fig. 162 apply strictly only to propellers with elliptical blades and hub diameter $\frac{2}{10}$ of propeller diameter, they are accurate enough for practical purposes for any other hub diameter likely to be found in practice and are reasonably good approximations for any blades of oval type.

5. Twisted Blades. — Propellers with detachable blades nearly always have them fitted so that they can be twisted slightly in the boss, thus increasing or decreasing the pitch. The blade flange holes are made oval, as shown in Fig. 156. The twist or rotation of the blade is about a line or axis through the center of the flange perpendicular to the shaft.

All pitch angles on the axis are changed a uniform amount.

For points of the blade away from the axis of twist the change is less, and for points of the helical surface a quarter of a revolution from the axis, if the surface were so great, there would be no change of pitch due to twist. For usual width of blade, however, the change in pitch angle is practically uniform over the blade and equal to the angle of twist. Hence the change of pitch due to twist will be investigated on this assumption.

Let y denote the diameter ratio, θ the pitch angle at a given point of radius r and pitch p. Let γ denote the angle of twist and y' the new diameter ratio after twisting.

Then $\tan \theta = \dfrac{p}{2\pi r} = \dfrac{1}{\pi y} \qquad y = \dfrac{1}{\pi} \cot \theta,$

$$\tan(\theta + \gamma) = \dfrac{1}{\pi y'},$$

$$y' = \dfrac{1}{\pi}\cot(\theta+\gamma) = \dfrac{1}{\pi}\dfrac{\cot\theta \cot\gamma - 1}{\cot\theta + \cot\gamma} = \dfrac{1}{\pi}\dfrac{\pi y \cot\gamma - 1}{\pi y + \cot\gamma} = \dfrac{y\cot\gamma - \dfrac{1}{\pi}}{\pi y + \cot\gamma}.$$

From the above formula, given y and γ, we can readily calculate y'. For a positive twist or value of γ the new diameter ratio is less than the old, the new pitch and pitch ratio being greater. For a negative twist the opposite holds.

The results are shown graphically in Figs. 163 and 164. In Fig. 163 the results are plotted upon diameter ratio. For each value of γ a curve is drawn showing the new values of diameter ratio plotted as ordinates over the old values as abscissæ. Contours are shown for each degree of positive and negative twist up to 6°.

Fig. 164 gives the same information as Fig. 163, but the results are plotted upon pitch ratio.

Figs. 163 and 164 illustrate the relative advantages and disadvantages of pitch ratio and diameter ratio when used as primary variables. Fig. 163 using diameter ratio, once the conception of diameter ratio is firmly grasped mentally, is simpler and more readily understood. This is largely because the diameter ratio at the tip of the blade is the natural starting point, and for any point of less radius the diameter ratio decreases directly as the radius. The conception of pitch ratio is more readily formed, but starting with the pitch ratio of the tips the pitch ratio increases inversely as the radius and becomes infinite for zero radius. In either case the tip value is a simple quantity of numerical value ranging in practice from .5 to 2. When using diameter ratio for any one blade the field covered, neglecting the hub, is that between zero and the tip value. When using pitch ratio the field is that between infinity and the tip value.

20. Theories of Propeller Action

1. Principles of Action Common to all Theories. — There have been a great many different theories of propeller action propounded, but none which has been generally accepted as agreeing fully with the facts of practical experience.

The principles underlying the chief English theories of propeller action are comparatively simple. The resulting formulæ are more or less complicated, but not difficult to apply. In any theory in connection with which mathematical methods are to be used it is almost necessary to regard the blade as having no thickness. Fig. 165, which partially reproduces Fig. 158, indicates the motion of a small elementary plane blade area of radius r, breadth dr, in a radial direction and circumferential length dl. Looking down we see this element with its center at O. If ω is the angular velocity of rotation of the shaft, the transverse velocity of the element is ωr. AOB is the pitch angle θ, BC the slip and BOC the slip angle φ. We know that $\tan \theta = \dfrac{p}{2\pi r}$. Considering Fig. 165 as a diagram of instantaneous velocities, the line OA or ωr represents the transverse velocity of the element. If there were no slip, the actual velocity would be parallel to OB since $BOA = \theta$. Then AB would denote the axial velocity.

$$AB = OA \tan \theta = \omega r \tan \theta = \omega r \dfrac{p}{2\pi r} = \dfrac{\omega p}{2\pi}.$$

When there is slip the transverse velocity of the element is unchanged, but the axial velocity is the speed of advance AC, which is denoted by V_A. BC is the slip and AB, the speed of the screw, is the same as the speed of advance when the slip is zero.

Denote the slip ratio by s.

Then $s = \dfrac{BC}{BA} = \dfrac{AB - AC}{AB} = \dfrac{\dfrac{\omega p}{2\pi} - V_A}{\dfrac{\omega p}{2\pi}} = \dfrac{\omega p - 2\pi V_A}{\omega p} = 1 - V_A \dfrac{2\pi}{\omega p}.$

Whence the speed of advance $V_A = \dfrac{\omega p}{2\pi}(1 - s) \quad BC = s\dfrac{\omega p}{2\pi}.$

If we take ω as angular velocity per second and r in feet, then OA or the transverse velocity is in feet per second, and hence all other velocities are in the same units.

Then we have

Velocity of blade element in the direction of the perpendicular to its plane $= CD = BC \cos \theta = s\dfrac{\omega p}{2\pi} \cos \theta.$

Axial or rearward component of above velocity $= CE = CD \cos \theta = s\dfrac{\omega p}{2\pi} \cos^2 \theta.$

Transverse component of above velocity $= DE = CD \sin \theta = s\dfrac{\omega p}{2\pi} \sin \theta \cos \theta.$

2. Three English Theories of Propeller Action. — There are three theories of propeller action whose detailed consideration will be of value. They are all contained in papers before the Institution of Naval Architects. The first was by Professor Rankine in 1865, the second by Mr. Wm. Froude in 1878 and the third by Professor Greenhill in 1888.

Rankine's fundamental assumption was that, as the propeller advanced with slip BC, all the water in an annular ring of radius r was given the velocity CD in a direction perpendicular to the face of the blade at that radius. Then, from the principle of momentum, the thrust from the elementary annular ring is proportional to the quantity of water acted upon in one second and to the sternward velocity EC communicated to it.

Froude considers the element as a small plane moving through the water along a line OC which makes a small angle ϕ with OB, the direction of the plane. Then Froude takes the normal pressure upon the elementary area which gives propulsive effect to vary as the area, as the square of its speed OC, and as the sine of ϕ the slip angle.

Greenhill makes a somewhat artificial assumption. He assumes that the propeller is working in a fixed closed end tube. The result is that the motion communicated to the water is wholly transverse and would be represented by CF in Fig. 165. The blade is first assumed smooth, so that the pressure produced by the reaction of the water is normal to the blade and has of course a fore and aft component which gives thrust. In all three theories the loss by friction is taken as that due to the friction of the propelling surface moving edgewise or nearly so through the water.

3. Relation between Direction of Pressure and Efficiency. — Neglecting friction for the present it is evident that all three theories start with a certain normal pressure. It follows that if this normal pressure be resolved into its axial and transverse components, say dT and dQ, we have

$$\frac{dT}{dQ} = \cot \theta = \frac{OA}{AB} = \frac{\omega r}{\frac{\omega p}{2\pi}} = \frac{2\pi r}{p}.$$

Hence $p\,dT = 2\pi r\,dQ$.

Now $2\pi r\,dQ$ = total work done during one revolution and hence, neglecting friction, $p\,dT$ = total work done during one revolution. Now the useful work $= dTp(1-s)$, hence the efficiency

$$e = \frac{dTp(1-s)}{dTp} = 1 - s.$$

It follows that, neglecting friction, if the reaction pressure from the water is normal to the face at all points of a screw of uniform pitch working with a slip s, the efficiency of each element and of the whole screw will be $1 - s$. Since the friction must reduce efficiency in all cases, it follows that upon the above supposition the efficiency of a screw cannot ever exceed $1 - s$. It is often thought that it is mechanically impossible for the efficiency of a screw to exceed $1 - s$. This, however, is not necessarily so. This limitation is associated with and dependent upon the assumption that the resultant pressure at each point of a screw surface is perpendicular to the surface. If the water can be made to move in such a manner that the resultant reaction is at an angle with the normal to the blade surface, we may have an efficiency, neglecting friction, greater than $1 - s$. This is an important point and worthy of careful investigation.

Referring to Fig. 166, suppose we have acting on a point O two forces OA and OB whose resultant OC makes an angle α with the axis of x, as indicated. Let the point O be moving with the velocity OE at the angle β with the axis of x as indicated. Then the work done by the reaction against the force $OA = OA \times OD$. The work done by the force $OB = OB \times ED = AC \times ED$.

Draw OF perpendicular to OC and denote EOF by γ. The ratio between the work done by the force OB and the work done by the reaction against OA is

$$\frac{AC \times ED}{OA \times OD} = \tan \alpha \quad \tan \beta = \frac{OD}{FD} \times \frac{ED}{OD} = \frac{ED}{FD}.$$

Now $\beta = 90° - \alpha - \gamma$.

The above is readily applied to the propeller problem. Referring to Fig. 167, which partially reproduces Fig. 166, consider an element at O whose pitch angle DOP is denoted by θ. Suppose OC is the resultant reaction upon the element O. Draw OF perpendicular to OC. Then AO is the transverse force upon the element denoted by q, say, while AC is the thrust denoted by t.

OD is transverse velocity V_t and DE is velocity of advance V_A. POD being the pitch angle θ, POE is the slip angle ϕ. Then the efficiency is the ratio between the useful work done by t and the gross input or work done by q and as before is $\dfrac{DE}{DF}$. Now if DE is speed of advance DP is speed of screw and $\dfrac{PE}{PD}$ = slip ratio = s. The efficiency of the element depends upon the directions of the resultant OC, and OF the perpendicular to it. Suppose the resultant OC is perpendicular to OP, then $\gamma = \phi$, F goes to P and the efficiency is $\dfrac{DE}{DP} = 1 - s$. It appears, then, to be rigidly demonstrable that if the resultant reaction at every point of a true screw is perpendicular to the face the efficiency of every element, and hence of the screw as a whole, is $1 - s$. As the direction of the resultant OC approaches the fore and aft line, or the perpendicular to AD, the efficiency of the element increases and would become unity if the resultant could become perpendicular to AD. As the direction of the resultant OC swings out from the fore and aft line beyond the perpendicular to the element, the efficiency becomes less than $1 - s$. Friction and head resistance always tend to swing the resultant in this direction, and the smaller the slip the smaller the values of AO and OC and the greater the relative effect of the force due to friction and head resistance.

I will now, neglecting friction at first, develop the formulæ for thrust and torque of a screw, following the three theories already referred to. For convenient comparison a uniform notation will be used, so far as practicable, differing slightly from the several notations of the original authors.

4. Rankine's Theory of Propeller Action. — Referring to Fig. 165 by Rankine's theory, considering the annular ring of mean radius r,

Annular area = $2\pi r dr$.

Volume of water acted on per second =

$$2\pi r dr \times AE = 2\pi r dr \times \frac{pR}{60}(1 - s\sin^2\theta).$$

Sternward velocity communicated $= EC = s\dfrac{\omega p}{2\pi}\cos^2\theta = s\dfrac{pR}{60}\cos^2\theta$.

Hence elementary thrust $=$ mass of water per second $\times$ sternward velocity imparted $= dT = \dfrac{w}{g} 2\pi r dr \dfrac{pR}{60}(1 - s\sin^2\theta)\, s\dfrac{pR}{60}\cos^2\theta$

$= \dfrac{w}{g}\dfrac{p^2 R^2}{3600} s(1 - s\sin^2\theta)\cos^2\theta\, 2\pi r dr.$

Let $q = \cot\theta = \dfrac{2\pi r}{p}$. Then $2\pi r dr = \dfrac{p^2 q}{2\pi} dq$. $\sin^2\theta = \dfrac{1}{1+q^2}$

$\cos^2\theta = \dfrac{q^2}{1+q^2}$.

Whence

$$dT = \dfrac{w}{g}\dfrac{p^2 R^2}{3600} s\left(\dfrac{q^2}{1+q^2} - s\dfrac{q^2}{(1+q^2)^2}\right)\dfrac{p^2}{2\pi} q dq$$

$$= \dfrac{w}{g}\dfrac{p^2 R^2}{3600}\dfrac{p^2}{2\pi} s\left\{ q dq - \dfrac{q dq}{1+q^2} - s\left(\dfrac{q dq}{1+q^2} - \dfrac{q dq}{(1+q^2)^2}\right)\right\}.$$

At the axis $q = 0$. Then, neglecting the hub, which a very slight investigation shows to have very little effect, if q denote now cotangent of the pitch angle of the blade *tips*, we have on integrating the expression for dT:

$$T = \dfrac{w}{g}\dfrac{p^2 R^2}{3600}\dfrac{p^2}{2\pi} s\left[\dfrac{q^2}{2} - \dfrac{\log_e(1+q^2)}{2} - s\left(\dfrac{\log_e(1+q^2)}{2} - \dfrac{1}{2}\dfrac{q^2}{1+q^2}\right)\right]$$

$$= \dfrac{w}{g}\dfrac{p^2 R^2}{3600}\dfrac{p^2 q^2}{4\pi} s\left[1 - \dfrac{\log_e(1+q^2)}{q^2} - s\left(\dfrac{\log_e(1+q^2)}{q^2} - \dfrac{1}{1+q^2}\right)\right].$$

Now $pq = 2\pi r$ $p^2 q^2 = 4\pi^2 r^2$ $\dfrac{p^2 q^2}{4\pi} = \pi r^2 = \dfrac{\pi d^2}{4}$ if d is extreme diameter. Whence

$$T = \dfrac{w}{g}\dfrac{p^2 R^2}{3600}\dfrac{\pi d^2}{4} s\left[1 - \dfrac{\log_e(1+q^2)}{q^2} - s\left(\dfrac{\log_e 1+q^2}{q^2} - \dfrac{1}{1+q^2}\right)\right]$$

Whence finally

$$T = \dfrac{\pi w}{14400\, g} p^2 d^2 R^2 s\left[1 - \dfrac{\log_e(1+q^2)}{q^2} - s\left(\dfrac{\log_e(1+q^2)}{q^2} - \dfrac{1}{1+q^2}\right)\right].$$

And the torque $Q = \dfrac{pT}{2\pi}$.

5. W. Froude's Theory of Propeller Action.

— Consider now Froude's theory.

If l is the total blade length of all blades at radius r, then the total elementary plane area at this radius is ldr. This area advances at the angle ϕ (Fig. 165), with velocity OC, and from Froude's experiments if a is a thrust coefficient, we have a resulting pressure normal to the blade $= ldr\, a\overline{OC}^2 \sin \phi$. The elementary thrust is equal to this pressure $\times \cos \theta$.

Then $dT = ldr\, a\overline{OC}^2 \sin \phi \cos \theta$.

Now
$$\overline{OC}^2 = \omega^2 r^2 + \frac{\omega^2 p^2}{4\pi^2}(1-s)^2 = \frac{\omega^2 p^2 q^2}{4\pi^2} + \frac{\omega^2 p^2}{4\pi^2}(1-s)^2$$
$$= \frac{p^2 R^2}{3600}\{q^2 + (1-s)^2\}.$$

$$\sin \phi = \frac{CD}{OC} = \frac{s\frac{\omega p}{2\pi}\cos \theta}{\frac{\omega p}{2\pi}\sqrt{q^2+(1-s)^2}} = \frac{s}{\sqrt{q^2+(1-s)^2}}\cos \theta.$$

Also $\cos^2 \theta = \dfrac{q^2}{1+q^2}$.

Whence
$$dT = ladrs\frac{p^2 R^2}{3600}\frac{q^2}{1+q^2}\sqrt{q^2+(1-s)^2}$$
$$= \frac{as}{3600}p^2 R^2 d^2 \frac{l}{d}\frac{q^2}{1+q^2}\sqrt{q^2+(1-s)^2}\frac{dr}{d},$$
$$= \frac{as}{3600}p^2 d^2 R^2 \frac{l}{d}\frac{p}{d}\frac{q^2}{1+q^2}\sqrt{q^2+(1-s)^2}\frac{dq}{2\pi}.$$

Whence, neglecting the hub as before,
$$T = \frac{a}{3600}p^3 R^2 ds \int_0^q \frac{l}{d}\frac{q^2}{1+q^2}\sqrt{q^2+(1-s)^2}\frac{dq}{2\pi}.$$

The quantity under the integral sign is evidently dependent only on shape and proportions of the propeller and independent of its dimensions. It can be determined in any case by graphic integration. For the present, let us denote it by the *symbol X*. Then from Froude's theory $T = \dfrac{a}{3600}p^3 R^2 ds X$, and as before $Q = \dfrac{pT}{2\pi}$.

6. Greenhill's Theory of Propeller Action.
— Coming finally to Greenhill's theory, we have (Fig. 165)

$$\text{Elementary area} = 2\pi r dr.$$

$$\text{Velocity of feed of the water} = AC = \frac{\omega p}{2\pi}(1-s) = \frac{pR}{60}(1-s).$$

$$\text{Transverse velocity} = s\frac{\omega p}{2\pi}\cot\theta = s\omega r = s\frac{2\pi R}{60}r.$$

$$\text{Transverse momentum per second} = \frac{w}{g} 2\pi r dr \frac{pR}{60}(1-s) s\frac{2\pi R}{60}r$$

$$= \frac{w}{g} p \frac{R^2}{3600} s(1-s) 4\pi^2 r^2 dr.$$

$$\text{Torque} = \text{transverse momentum} \times r.$$

Whence $dQ = \dfrac{w}{g} p \dfrac{R^2}{3600} s(1-s) 4\pi^2 r^3 dr.$

$$dT = \frac{2\pi dQ}{p} = \frac{w}{g}\frac{R^2}{3600}s(1-s) 8\pi^3 r^3 dr.$$

Integrating from $r = 0$ to $r = \dfrac{d}{2}$ we have

$$T = \frac{w}{g}\frac{R^2}{3600}s(1-s) 2\pi^3 r^4 = \frac{\pi^3 w}{28800\, g} d^4 R^2 s(1-s).$$

And as before $Q = \dfrac{pT}{2\pi}.$

In connection with Greenhill's theory, it should be pointed out that the excess pressure at any radius is very simply expressed. We have above $dT = \dfrac{w}{g}\dfrac{R^2}{3600} s(1-s) 8\pi^3 r^3 dr.$

But if ΔP be the excess pressure per unit area, $dT = 2\pi r dr \Delta P.$

Whence dividing through $\Delta P = \dfrac{w}{g}\dfrac{R^2}{3600} s(1-s) 4\pi^2 r^2.$

In other words, the excess of pressure varies as the square of the radial distance from the axis.

7. Comparison of Theories with Each Other.
— Now, comparing the three formulæ for thrust and torque, it is seen that each one is composed of a coefficient, of a term involving the

dimensions and revolutions or speed, and of a term varying with shape, proportions and slip but independent of the dimensions. Assuming, as is evidently possible, that we can expand X in the formula from Froude's theory in the form $\alpha - \beta s +$ negligible terms, we can write for each formula $T = \phi(pdR)(\alpha s - \beta s^2)$.

For Froude's theory $\phi(pdR) = p^3 dR^2$ and for Rankine's theory $\phi(pdR)$ is $p^2 d^2 R^2$. For Greenhill's theory $\phi(pdR)$ is $d^4 R^2$, of the same dimensions as before but independent of the pitch. Now, considering α and β, it is evident that by the formula for Froude's theory β will be very small indeed compared with α. In the Rankine theory formula β will be smaller than α, but relatively larger than in the Froude theory formula. In the Greenhill theory formula $\beta = \alpha$ always.

Still neglecting friction, we would have on the theory of all motion communicated to the water perpendicular to the blade

$$Q = \frac{pT}{2\pi} = \frac{p}{2\pi} \phi(pdR)(\alpha s - \beta s^2).$$

As a matter of fact, a very brief examination of experimental results shows that this cannot hold. If it were true, we could never have an efficiency greater than $1 - s$, and even when friction is considered we get experimental efficiencies greater than $1 - s$. So it appears well to adopt tentatively as the general expression for the torque $Q = \frac{p}{2\pi} \phi(pdR)(\gamma s - \delta s^2)$.

8. Friction and Head Resistance. — Now consider friction and head resistance. Referring to Fig. 165, if f denote the coefficient of friction and dA an elementary area, we have with close approximation frictional resistance $= f dA \overline{OB}^2$. In practice ϕ is a much smaller angle than indicated in Fig. 165, where it is exaggerated for clearness. Suppose f is large enough to cover all edgewise resistance — skin friction and head resistance together.

Then $dA = ldr$, $\overline{OB}^2 = p^2 R^2 \cosec^2 \theta = p^2 R^2 (1 + q^2)$, $q = \frac{2\pi r}{p}$, $dr = \frac{p}{2\pi} dq$.

Then $F = f \frac{dl}{d} \frac{p}{2\pi} dq p^2 R^2 (1 + q^2) = f p^2 d R^2 \frac{p}{2\pi} \left\{ \frac{l}{d} (1 + q^2) dq \right\}$.

Fore and aft component = Deduction from thrust

$$= F \sin \theta = \int p^2 dR^2 \frac{p}{2\pi} \frac{l}{d} \sqrt{1+q^2} dq = dT_f.$$

Transverse component $= F \cos \theta = \int p^2 dR^2 \frac{p}{2\pi} \frac{l}{d} q \sqrt{1+q^2} dq.$

Difference of torque $= F \cos \theta \times r = F \cos \theta \frac{pq}{2\pi}$

$$= \int p^2 dR^2 \frac{p^2}{4\pi^2} \left(\frac{l}{d} q^2 \sqrt{1+q^2} dq \right) = dQ_f.$$

Integrating,

Deduction from thrust for friction

$$= T_f = \int p^3 dR^2 \int \frac{l\sqrt{1+q^2}}{2\pi d} dq = \int p^3 dR^2 Y, \text{ where } Y = \int \frac{l\sqrt{1+q^2}}{2\pi d} dq.$$

Addition to torque for friction $= Q_f = \frac{p}{2\pi} \int p^3 dR^2 \int \frac{q^2 l \sqrt{1+q^2}}{2\pi d} dq$

$$= \frac{p}{2\pi} \int p^3 dR^2 Z, \text{ where } Z = \int \frac{l}{2\pi d} q^2 \sqrt{1+q^2} dq.$$

Since for the working portions of actual propellers q is greater than 1, we will have in practice Z much greater than Y, and it is reasonable to ascribe the total friction loss to increase of torque. If we assume $\frac{l}{d}$ constant = mean width ratio × number of blades, we can readily determine a curve of Z on q by plotting a curve of $\frac{q^2 \sqrt{1+q^2}}{2\pi}$ and integrating graphically.

For actual propellers Y and Z can be determined without difficulty by plotting on q curves of $\frac{l}{2\pi d} \sqrt{1+q^2}$ and $\frac{l}{2\pi d} q^2 \sqrt{1+q^2}$ and integrating graphically.

Fig. 168 shows curves of Y and Z and of $\frac{Y}{Z}$ for elliptical blades with hub diameter .2 the extreme diameter, plotted upon pitch ratio, and Fig. 169 shows curves of X for various values of s, namely, $s = 0$, .20, and .40.

9. Final Formulæ on Theories of Rankine, Froude and Greenhill. — Then the final formulæ for thrust and torque including the friction term can be expressed in the forms below:

Rankine's Theory: $T = p^2 d^2 R^2 (\alpha s - \beta s^2) - f d p^3 R^2 Y,$

$$Q = \frac{p}{2\pi} [p^2 d^2 R^2 (\gamma s - \delta s^2) + f d p^3 R^2 Z].$$

Froude's Theory: $T = p^3 d R^2 (\alpha s - \beta s^2) - f d p^3 R^2 Y,$

$$Q = \frac{p}{2\pi} [p^3 d R^2 (\gamma s - \delta s^2) + f d p^3 R^2 Z].$$

Greenhill's Theory: $T = d^4 R^2 (\alpha s - \beta s^2) - f d p^3 R^2 Y,$

$$Q = \frac{p}{2\pi} [d^4 R^2 (\gamma s - \delta s^2) + f d p^3 R^2 Z].$$

The above equations are simply to show the form of the expressions. They do not imply that α and β in the Rankine Theory equation will be the same as in the Froude or Greenhill Theory equation, but simply that in each case α and β will be constant for a given propeller. The actual values of the constants will vary with the theory used.

The formulæ on Froude's theory are expressed in the above form, as previously noted, by assuming that X can be expanded with sufficient approximation in the form $C - sD$, where C and D are independent of s. It is evident from Fig. 169 that this can be done and that D is much smaller than C.

In all the theories, as has already been pointed out, it is assumed that the net reaction at each point is perpendicular to the blade surface. If this were true, we would always have $\alpha = \gamma$, $\beta = \delta$, and the efficiency could never exceed $1 - s$ even if there were no friction. Since experience shows this is not the case, and as from considering the probable motion of a particle of water it is evidently not necessary that the net momentum impressed upon it shall be perpendicular to the blade surface, I have, while following the same form, used different coefficients for the torque expression, expecting that these coefficients γ and δ need not necessarily be the same as α and β used for thrust.

It seems difficult at first sight to conceive of any fluid action upon a frictionless surface that is not at right angles to it, but if we consider the matter from the point of view of the velocity impressed upon the water the difficulty disappears. The suction of the propeller upon the water ahead of it causes a velocity which is practically all axial, or in the direction perpendicular to the plane of the propeller disc. Hence, the reaction upon the water is partly axial before the water reaches the propeller disc and partly normal or nearly so as the water passes through the disc, the final resultant being at an angle with the normal in the direction which we have seen tends to make the efficiency greater than $1 - s$.

10. Comparison of Theories with Facts of Experience. — It does not require much reflection to render it evident that none of the three theories considered correctly represents the physical phenomena. This conclusion is very strongly confirmed by the results of model experiment and general experience.

On Rankine's theory the water while passing through the screw disc is given the sternward velocity EC (Fig. 165). This can occur only if the stream contracts materially while passing through the propeller or if a material quantity of water from abreast the disc is always flowing into it. Neither motion seems reasonable. Furthermore, on Rankine's theory, the thrust and torque are independent of the blade surface, one assumption of Rankine's theory being that "the length of the screw and number of its blades are supposed to be adjusted by the rules deduced from practical experience, so that the whole cylinder of water in which the screw revolves shall form a stream flowing aft."

Practical experience with model propellers shows clearly that the result assumed by Rankine is unattainable. Rankine's theory further ignores variations of pressure which must occur in propeller action.

Froude's theory goes to the opposite extreme of Rankine's. It assumes that the thrust increases always in direct ratio to the area. Model experiments show conclusively that, while within practicable limits thrust does increase as long as area increases, the increase in thrust is by no means proportional to the area increase, the rate of increase with area diminishing steadily as area increases.

Greenhill's theory has the same obvious defect as Rankine's, in that it neglects the effect of area of blade. The portion I have used ignores the sternward velocity, deducing thrust entirely from the pressure set up by rotating the water in the disc, but it should be pointed out that his 1888 paper gives some consideration to other possible motions involving axial velocity of slip in the water.

As, then, it seems that no theory we have considered can exactly represent the action of propellers, it would be necessary, in case we wished to adhere to formulæ, to compare each formula with experimental results and select that one which seemed to agree most closely. Then using this as a semi-empirical formula, with coefficients and constants deduced from experiments or experience, problems could be satisfactorily dealt with. But it will be observed that each formula is of the proper dimensions to satisfy the Law of Comparison. Hence if either formula holds, the Law of Comparison will hold, and experimental results, instead of being utilized to supply coefficients and constants for use with a formula, can be reduced to a form to be utilized directly by graphic methods. Per contra, if the Law of Comparison does not hold, the formulæ on all of the three theories will fail. In either case there is obviously no advantage from a practical point of view in attempting to reduce the formulæ to forms for use in practice. A serious practical disadvantage is the fact that the formulæ use a true slip, based upon true pitch, or a blade of no thickness. The face pitch of a blade with thickness, or its nominal pitch as it may conveniently be called, is very different from the virtual or effective pitch, and this fact causes material complications in using formulæ.

11. Slip Angle Values. — In connection with theories of propeller action it is desired to invite particular attention to the fact that propellers in practice operate with slip angles that are very small indeed. A slip of 20 per cent somehow seems to imply a large angle, but as a matter of fact it usually means in practice an angle of from $2\frac{1}{2}$ to 5 degrees only, and most propellers show their maximum efficiency at slips below 20 per cent.

Referring to Fig. 165, where ϕ denotes the slip angle, we have

$$\sin \phi = \frac{CD}{CO} = \frac{BC \cos \theta}{\sqrt{OA^2 + AC^2}}$$

$$= \frac{s \dfrac{\omega p}{2\pi} \cos \theta}{\sqrt{\omega^2 r^2 + \dfrac{\omega^2 p^2}{4\pi^2}(1-s)^2}} = \frac{s \dfrac{\omega p}{2\pi} \dfrac{\omega r}{\sqrt{\omega^2 r^2 + \dfrac{\omega^2 p^2}{4\pi^2}}}}{\sqrt{\omega^2 r^2 + \dfrac{\omega^2 p^2}{4\pi^2}(1-s)^2}}.$$

Let y denote diameter ratio $= \dfrac{d}{p} = \dfrac{2r}{p}$. Then $r = \dfrac{py}{2}$.

Substituting, clearing and reducing, we have finally

$$\sin \phi = s \frac{\pi y}{\sqrt{1 + \pi^2 y^2} \sqrt{\pi^2 y^2 + (1-s)^2}}.$$

Hence given s and y the value of ϕ is fixed.

Fig. 170 shows graphically the relation between slip angle ϕ, slip s and diameter ratio. Also at the top of the figure is a scale for pitch ratio, but reference to diameter ratio is more illuminating. Considering a screw of uniform face pitch it is seen that for a given slip per cent the slip angle is a minimum where the diameter ratio is greatest — at the blade tip. As we go in from the tip the slip angle increases, reaching a maximum when diameter ratio = .3 about, and then rapidly decreasing to zero at the axis. But on account of the hub the falling off of slip angle below diameter ratio of .3 is immaterial, and to all intents and purposes slip angle increases from tip to hub. The actual values for the diameter ratios and slips found in practice say below diameter ratio of 1.1 and slip ratio of .30 are quite small.

The maximum efficiency of most propellers corresponds to a nominal slip in the neighborhood of 15 per cent, and for this the maximum slip angle at the hub is less than 5° and for the most important part of the blade it is in the vicinity of 3°. These are small angles, and the fact that slip angles are so small should never be lost sight of in considering operation of propellers.

21. Law of Comparison Applied to Propellers

1. Formulæ for Applying Law of Comparison to Propellers. — In connection with the Law of Comparison the formulæ for the application of the law to propellers have been already indicated, but they are recapitulated below.

Suppose we have a propeller and a smaller similar propeller or model. Let us use symbols as in the table following:

	For Large Propeller.	For small Propeller or Model.
Diameter in feet.	D	d
Revolutions per minute.	R	r
Speed of advance in knots.	V	v
Thrust in pounds.	T	t
Torque in pound-feet.	Q	q
Pressure on propeller, pounds per square inch.	P_1	p_1
Power absorbed.	P	p

Then if λ denote the ratio of linear dimensions of model and full-sized screw we have the following relation:

$$D = \lambda d, \quad R = \frac{r}{\sqrt{\lambda}}, \quad V = v\sqrt{\lambda}, \quad T = \lambda^3 t,$$

$$Q = \lambda^4 q, \quad P_1 = \lambda p_1, \quad P = \lambda^{3.5} p.$$

2. Conditions Governing Application of Law of Comparison. — Note that for the complete applicability of the Law of Comparison $P_1 = \lambda p_1$, or all pressures should be in the ratio of the linear dimensions. Now the pressure under which a model propeller works is made up of two components — the water pressure due to its submersion and the constant pressure of the atmosphere exerted upon the surface and transmitted through the water.

When we consider the full-sized propeller we find the pressure due to submersion is or readily can be increased to scale; but the atmospheric pressure is not increased, and hence this component of the total pressure upon the full-sized propeller is only $1 \div \lambda$ of the value needed to have the Law of Comparison exactly applicable. Hence it might be inferred that, as the conditions required by the Law of Comparison are not present, model experiments are of little value in the investigation of propellers. But upon consideration it is evident that in each case the atmospheric pressure is transmitted through the water, appearing both in front of and behind model and propeller; and, since the forces upon model and propeller are due to reactions caused by the motions impressed upon the water, the Law of Comparison will apply provided the motions of the water around model and propeller are similar. The pressure relation fails in precisely the same way in passing from models to ships, but in this case the motions produced are not affected by the surface pressure and the Law of Comparison holds. Hence we may rely upon the Law of Comparison and design propellers upon the basis of model results if we can but be sure that the motions of the water around model and propeller will be similar.

Now, we are reasonably certain that until we reach speeds and thrusts at which the phenomenon known as cavitation makes its appearance the motions of the water around model and propeller are so nearly similar that the Law of Comparison is applicable. When cavitation is present the Law of Comparison fails, because, as will be seen when discussing cavitation, the model does not cavitate as a rule, and hence results from it are an unsafe guide when dealing with the full-sized screw. But the majority of propellers as fitted are not very seriously, if at all, interfered with by cavitation, and for such propellers model experiments are of great value, since the Law of Comparison may be somewhat confidently relied upon in connection with them. Exact comparison of experimental data from a model and a full-sized propeller of large dimensions has never been made, but experiments at the United States Model Basin showed that for small or model propellers ranging from 8 inches to 24 inches in diameter the Law of Comparison applies reasonably well. (See paper entitled "Model Basin Gleanings," Transactions Society of Naval Architects and Marine Engineers for 1906.)

We have seen that theoretical formulæ for propeller action all give the result that for a given propeller form advancing with a given slip the thrust and torque vary as the square of the speed of advance and also, that the thrust varies as the square and the torque as the cube of the linear dimensions.

If this is the case, the Law of Comparison necessarily holds. There are a number of reasons for thinking that thrust and torque for a given propeller advancing with given slip vary as the square of the speed of advance. If the lines of flow or paths followed by the particles of water are the same, whatever the speed, then thrust and torque must vary as the square of the speed. For then the quantity of water acted upon must vary directly as the speed, and the velocity communicated to each particle acted on must vary directly as the speed. Hence the momentum generated per second, to which thrust and torque are proportional, must vary as the square of the speed.

Experiments made at the United States Model Basin in 1904 with 16-inch model propellers between speeds of three and seven knots showed that within the limits of experimental error thrust and torque varied very approximately as the square of the speed. The propellers whose thrust varied as a greater power of the speed than the square were usually those with very narrow blades. Those whose thrust varied as a lesser power of the speed than the square were usually those with very broad blades.

Finally, experience in analyzing accurate trial results shows that, broadly speaking, when cavitation is not present, at speeds where the resistance of the ship is varying as the square of the speed the slip is practically constant, which of course means that the thrust of the propeller advancing with this constant slip varies as the square of the speed.

At speeds for which the resistance of the ship is varying as a less power of the speed than the square the slip is falling off, and at speeds for which the resistance is varying as a greater power of the speed than the square the slip is increasing. This is fairly strong evidence from accumulated experience that the thrust of full-sized propellers varies as the square of the speed of advance.

In the light of present knowledge we appear to be warranted in concluding that the Law of Comparison applies to propeller action sufficiently well for practical purposes until cavitation appears. There is reason to believe, however, that cases have occurred where cavitation has been present without being suspected.

22. Ideal Propeller Efficiency

1. Thrust, Power and Efficiency of Ideal Propelling Apparatus. — In a paper before the Society of Naval Architects and Marine Engineers, in 1906, entitled "The Limit of Propeller Efficiency," Assistant Naval Constructor W. McEntee, without setting up any special theory of propeller action, has pointed out the limit of propeller efficiency beyond which we cannot go.

Suppose we have a frictionless propelling apparatus discharging a column of water of A square feet area directly aft with an absolute velocity u, while the speed of the ship is v, both v and u being measured in feet per second. Then if w denote the weight per cubic foot of the water, the weight acted on per second is $wA(v+u)$ and the mass is $\frac{w}{g}A(v+u)$.

The reaction or thrust $T = \frac{w}{g}A(v+u)u$ being equal to the sternward momentum generated per second.

Useful work $= \frac{w}{g}A(v+u)vu$.

There being no friction, the lost work is simply the kinetic energy in the water discharged. Hence we have

Lost work $= \frac{w}{g}A(v+u)\frac{u^2}{2}$.

Gross work $= \frac{w}{g}A(v+u)vu + \frac{w}{g}A(v+u)\frac{u^2}{2}$.

Efficiency $e = \dfrac{\text{Useful work}}{\text{Gross work}} = \dfrac{v}{v+\frac{u}{2}}$.

Also solving for u in the equation for thrust T, we get

$$u = \sqrt{\frac{v^2}{4}+\frac{gT}{wA}} - \frac{v}{2}.$$

Substituting in the expression for efficiency, we have

$$e = \frac{4}{3+\sqrt{\dfrac{4g}{w}\dfrac{T}{Av^2}+1}}.$$

This expression for maximum efficiency must involve the assumption that the water is discharged without increase of pressure. The effect of an increase of pressure would be to decrease the efficiency, since work done against pressure would be done with efficiency $\frac{v}{v+u}$. Hence we conclude that the value of e above is the maximum that could be attained by a perfect propeller. Suppose, applying this to a screw propeller, we write $\frac{\pi d^2}{4}$ for A, where d is the diameter of the propeller in feet. Now if U denote useful horse-power delivered by the propeller and P denote gross horse-power, or horse-power delivered to the propeller, we have $eP = U = \frac{Tv}{550}$, whence $T = \frac{550\, eP}{v}$. Also $v = \frac{6080}{3600} V$, where V is speed of advance in knots. And $g = 32.16$, $w = 64$ for sea water. Substituting and reducing, we have finally

$$\frac{P}{d^2 V^3} = \frac{1}{292.2} \times \frac{16 - 24\, e + 8\, e^2}{e^3} = \frac{2 - 3\, e + e^2}{36.52\, e^3}.$$

2. Discussion of Ideal Efficiency Results. — From the above, Figs. 171 and 172 were drawn, Fig. 171 showing contours of efficiency on values of V as abscissæ and of $\frac{P}{d^2}$ as ordinates and Fig. 172 showing contours of efficiency on values of d as abscissæ and of $\frac{P}{V^3}$ as ordinates.

These figures should not be mistaken as representing actual efficiencies that are attainable. They are purely ideal diagrams, and their indication that efficiency always increases with increase of diameter is misleading if followed too far as regards actual propellers. They are interesting and instructive, however, as giving us in any particular case a limiting efficiency beyond which we could not possibly go and which we must fall short of in practice.

In Fig. 171 there is shown a supplementary scale of $\frac{P}{A}$, or power per square foot of disc area. This of course bears a constant ratio to $\frac{P}{d^2}$.

A striking result of the formula for ideal propeller efficiency is the high efficiency attained with large slips. The expression for slip ratio s_1, in terms of v and u, is $s_1 = \dfrac{u}{v+u}$. The formula for efficiency is $e = \dfrac{v}{v + \dfrac{u}{2}}$, whence expressing e in terms of s_1, we have

$$e = \frac{1 - s_1}{1 - \dfrac{s_1}{2}}.$$

Fig. 173 shows a curve of e plotted on s_1, as deduced from the above formula. This efficiency is everywhere above the line $1 - s_1$.

In this connection it is interesting to recall that numerous experiments with model propellers at high slips show an efficiency greater than $1 - s$. It should be remembered, however, that in the case of these actual small propellers s is derived from the pitch of the driving face, while in the ideal formula s_1 is based upon the assumed sternward velocity u of the water, and the water is not supposed to have any transverse velocity. The actual sternward velocity of the water in the operation of actual propellers is not easy to determine or estimate, and transverse velocity is always present.

On Rankine's theory we can readily establish the relation between s and sternward velocity. In Fig. 165 the sternward velocity is $EC = s \cos^2 \theta$. This is much less than BC, the slip velocity. While we cannot say that in actual cases the sternward velocity is EC, there is no question that it is very much less than BC, the slip velocity. It could be equal to BC only if there were no transverse velocity communicated to the water, and there is no question that in practice transverse velocity is always communicated. A very common mistake is to consider the sternward velocity communicated to the water the same as the slip velocity, or BC in Fig. 165.

23. Model Experiments — Methods and Plotting Results

1. **Experimental Propeller Models and Testing Methods.** — Having concluded that the Law of Comparison is applicable to many cases of propeller action so that experiments with model propellers may be expected to be of value, I will now go into this

question. Numerous experiments with model propellers have been made at the United States Model Basin. The details of the apparatus and methods used will be found in the author's paper of 1904 before the Society of Naval Architects and Marine Engineers entitled "Some Recent Experiments at the United States Model Basin." The experimental gear described in that paper has been changed subsequently only in minor details as improvements suggested themselves.

The model propellers are usually made of composition, accurately finished to scale. Most of them have been 16 inches in diameter. When being tested the model propeller is attached to a horizontal shaft projecting ahead of a small boat which is rigidly secured to the carriage traversing the basin. The shaft projects so far that the propeller is practically unaffected by the presence of the following boat. The propeller shaft center is 16 inches below the surface of the water, so that the blade tips of a 16-inch model are immersed 8 inches, or one-half of a diameter. The hub is fitted with fair-waters in front and behind. Fig. 174 shows the arrangement for a hub $3\frac{1}{8}$ inches in diameter, which was a standard hub diameter adopted for all models which did not represent actual propellers. For models of actual propellers, the hubs represent to scale the actual hubs, appropriate fair-waters being fitted.

Dynamometric apparatus, described in detail in the paper above referred to, enabled the torque and thrust of the model propeller to be accurately determined.

By making runs with dummy hubs having no blades attached the hub effect was eliminated as far as possible, the endeavor being to determine experimentally the torque and thrust of the blades alone.

The greater number of experiments were made at a 5-knot speed of carriage, this speed of advance being kept constant as nearly as possible, and slip being varied by varying the revolutions of the propeller. In the early stages of the experiments, however, a number of propellers were tested at speeds of advance ranging from 3 knots to 7 knots, and between these speeds it was found that within the limits of error the thrust and torque at constant slip varied practically as the square of the speed. As has been already pointed out, this agrees with the formulæ of Rankine, Froude and Greenhill, which agree in making thrust and torque vary as R^2, and when slip is constant the speed of advance varies as R.

In making the 5-knot experiments speeds of individual runs of a series would differ slightly from 5 knots, and the thrust and torque were reduced to the 5-knot speed by taking them to vary as the square of the speed.

2. Methods of Recording Experimental Results. — As will be seen upon consulting the original paper, during a run the thrust and torque are recorded continuously, and after uniform conditions have been reached the time and revolutions are recorded every 32 feet. For convenience the thrust and torque at 5 knots speed are plotted initially upon the revolutions made by the propeller upon a 64-foot interval — denoted by ρ_1, which is one of the quantities observed.

Fig. 176 shows curves of thrust and torque plotted thus for the model propeller whose developed blade outline and blade sections are shown in Fig. 175. This is a 16-inch three-bladed model propeller of the true screw, ordinary type, the pitch being 16 inches — pitch ratio 1.00 — the blades being elliptical, of .25 mean width ratio, and the sections ogival. The hub diameter is .2 the propeller diameter. The curves of Fig. 176 are plotted upon ρ_1, or revolutions per 64-foot interval. Lines showing the values of ρ_1 for various values of the slip are shown on the figure, the slip being based upon the nominal pitch of 16.0 inches. These lines are not equally spaced, for, p denoting pitch in feet and s the slip ratio, we have $p\rho_1(1-s) = 64$, or $\rho_1 = \dfrac{64}{p(1-s)}$. For equal increments of s the interval between successive corresponding values of ρ_1 constantly increases.

It will be observed that ρ_1 is dependent upon the pitch and slip only and for a given slip is quite independent of the speed. Furthermore, the experimental apparatus was such that ρ_1 was determined with great accuracy. Thus it was a very suitable quantity to use as a primary variable upon which to plot the experimental values of thrust and torque for the purpose of deducing curves of the same.

24. Model Propeller Experiments — Analysis of Results

1. Methods of Plotting Information Derived from Experiment.
— The results of model experiments having been plotted as curves of thrust and torque upon the revolutions made upon a 64-foot length as shown in Fig. 176, the lines for various definite nominal slips being indicated upon the same diagram, the subsequent treatment depends upon the purpose in view.

For purposes of analysis, comparison of efficiency, etc., the methods would naturally differ from those most convenient for use in design.

When we consider the best method of plotting for purposes of analysis, etc., curves deduced from model propeller experiments, it soon becomes evident that we may with advantage record the data as curves of coefficients — quantities that do not vary with dimensions. As abscissæ for such curves the slip ratio is a desirable quantity to use. It is not dependent upon size or speed, and is one of the primary variables involved in screw action.

2. Virtual and Nominal Pitch and Slip. — The question at once arises, however, whether we should use nominal slip, namely, slip based upon the pitch of the screw face, or real slip, i.e., slip based upon the virtual pitch, or pitch of the ideal blade of no thickness which would act as the actual blade.

This virtual pitch is a thing very different from the nominal pitch. The ignoring of this fact has had a great deal to do with the prevention of correct conclusions as to propeller performance. In the case of a true screw the pitch of the driving face is known, but every point of the back has a pitch, and the back has much to do with screw performance. One might think without looking into it that for ordinary cases the pitch of the back is nearly the same as that of the face. The truth is that the pitch of the back varies prodigiously from the pitch of the face. Fig. 175 shows blade sections of a screw of not unusual blade thickness and of face pitch equal to diameter, the sections being of the usual ogival type. Taking face pitch and diameter as 16 feet, Fig. 177 shows plotted on radius the pitch of the back at the leading edge and at the following edge. It is seen that the pitch of the leading portion of the back will average somewhere about 50 per cent less than the uniform pitch of the face or the nominal pitch. On the other hand, the pitch of the following edge of the back is on the average somewhat more than 50 per cent greater than the nominal pitch. It is quite obvious that such a screw cannot act as a theoretical screw, having blades of no thickness and of the uniform pitch of the face. It is evidently desirable to find some method of determining for a known screw its virtual pitch, or equivalent uniform pitch. Now, for all formulæ we have, neglecting friction, no thrust or torque at zero slip. Experimental results with screws of uniform nominal pitch and ogival type of blade section always show as in Fig. 176 both thrust and torque when the slip calculated on the nominal pitch is zero. It follows that for such screws the virtual pitch is greater than the nominal pitch. This might be inferred, too, from the fact that at the rear of the blade the pitch of the back is always greater than the nominal, and, if the back has any influence at all, it must increase the virtual pitch over the nominal pitch. Suppose, now, we consider some experimental results. Fig. 178 shows upon an enlarged scale the lower part of Fig. 176, being curves of thrust and torque as determined experimentally for a 16-inch model of the propeller of Fig. 175 plotted upon p_1, or revolutions required to traverse a distance of 64 feet, the speed of advance of the propeller being kept constant at 5 knots. Now on any theory we have at true zero slip a negative thrust T_f and a positive torque Q_f, both being due to the friction and head resistance only. From the formulæ given when considering the theories of Rankine, Froude and Greenhill,

$$T_f = -\int dp^3 R^2 Y, \qquad Q_f = \frac{p}{2\pi} \int dp^3 R^2 Z.$$

Whence $\int dp^3 R^2 = -\dfrac{T_f}{Y} = \dfrac{2\pi Q_f}{pZ}.$

Whence $\dfrac{pT_f}{2\pi Q_f} = -\dfrac{Y}{Z} = \dfrac{pT}{2\pi Q}$ when $s = 0$.

Now $\dfrac{Y}{Z}$ is a fixed quantity for the propeller. For the propeller in question it is equal to .236, from Fig. 168. Fig. 178 shows the method to be followed. If $s = 0$, we have $p = \dfrac{64}{\rho_1}$. So we can plot a curve of p on ρ_1 as abscissa. Also we can plot the curve of $\dfrac{pT}{2\pi Q}$, as shown. This has the value $-.236 = -\dfrac{Y}{Z}$ at $\rho_1 = 42.11$, for which $p = 1.520$ feet. Then from the diagram the virtual pitch of the screw is 1.520 feet, or 18.24 inches, or 1.140 times the nominal pitch of 16 inches. Very frequently the virtual pitch is taken such that zero slip will give zero thrust. This is not quite correct, however, because at zero thrust there is a small negative thrust due to friction and an equal and opposite positive thrust due to slip. The error, however, is not great. In Fig. 178, at zero thrust $\rho_1 = 42.70$, $p = 1.499$ feet $= 17.99$ inches. The difference in virtual pitch is only about $1\frac{1}{2}$ per cent, and as it is very difficult to make model propeller experiments with minute accuracy, it is hardly worth while in practice to use the exact method. Moreover, while we should always bear in mind that the nominal pitch is not the real pitch or virtual pitch, it is very desirable to use always the nominal pitch in practical cases. We shall see that this can be done, so that the question of virtual pitch, though of great scientific interest, is academic rather than practical. So, except for special applications, results for true screws of uniform face pitch will be plotted upon nominal slip corresponding to the face or nominal pitch.

3. Determination of Efficiency. — The ordinates for the curve of efficiency plotted upon nominal slip are readily and simply determined from the curves of thrust T in pounds and torque Q in pound-feet. For if p denote pitch in feet, R revolutions per minute and s the slip, speed of advance is $p(1-s)R$, and useful work done in a minute $= TpR(1-s)$. The gross work, or work delivered to the model, is $Q \times 2\pi R$.

Now efficiency $=$ (Useful Work) $\div$ (Gross Work) $= \dfrac{TpR(1-s)}{2Q\pi R}$

$= \dfrac{T}{Q} \dfrac{p(1-s)}{2\pi}$.

Note that the quantity $p(1-s)$ is the advance of the screw

for one revolution, and its value is the same whether nominal or virtual pitch is used, the slip in each case being that appropriate to the pitch. Since at the Model Basin the curves of T and Q are plotted upon ρ_1, the revolutions per 64-foot interval, it is convenient to use this in the efficiency formula.

We have $\rho_1 = \dfrac{64}{p(1-s)}$. Substituting and reducing, we have finally

$$e = \frac{10.186}{\rho_1} \frac{T}{Q}.$$

The values of T, Q and p being taken off for the values of ρ_1 for the various slips, as indicated in Fig. 176, the efficiencies are readily calculated and plotted on slip.

4. Characteristic Coefficients. — The next question is as to the curves of coefficients which will completely characterize the propeller. Various coefficients may be used. Papers by the author and Messrs. Curtis and Hewins of the Model Basin staff before the Society of Naval Architects and Marine Engineers give various forms of coefficients, but it is believed that those given below are simple and convenient.

We have to deal with the power absorbed or propeller power P, the useful or net power E, the speed of advance in knots V, the revolutions per minute R, the slip and the size.

Whatever formula we use we are led to the same type of expression connecting power absorbed, speed of advance and diameter. Thus using Rankine's formula,

$$Q = \frac{p}{2\pi}(p^2 d^2 R^2(\gamma s - \delta s^2) + f p^3 d R^2 Z).$$

Gross power $\quad P = \dfrac{2\pi QR}{33000} = \dfrac{1}{33000}(p^3 R^3 d^2(\gamma s - \delta s^2) + f p^3 R^3 p d Z).$

Now $\quad pR = \dfrac{101\frac{1}{3} V}{1-s}.\quad$ Let $p = \dfrac{d}{m}.\quad pd = \dfrac{d^2}{m}.$

Then $\quad P = d^2 V^3 \left[\dfrac{(\gamma s - \delta s^2)(101\frac{1}{3})^3}{(1-s)^3 \times 33000} - \dfrac{fZ}{m}\dfrac{(101\frac{1}{3})^3}{(1-s)^3 \times 33000} \right].$

Using either Froude's or Greenhill's formula we are led to the same expression except that the Froude theory formula will have the first term in the parentheses divided by m and the Greenhill theory formula will have it multiplied by m^2. In either case we may write $P = \dfrac{d^2V^3A}{1000}$, where A is a coefficient independent of the size and speed of the screw but varying with the slip and dependent upon shape, proportions, etc. The divisor 1000 is introduced simply in order to give A a greater value than unity in practical cases. Otherwise A would be inconveniently small. Evidently then a curve of A plotted on slip will completely characterize a screw as regards the important question of its capacity to absorb power.

If E denote the useful or effective horse-power delivered by the screw, we have $E = eP = d^2V^3\dfrac{eA}{1000}$.

Let us denote eA by B. Then curves of e, A and B plotted upon slip will completely characterize the action of a propeller of given features independently of size and speed.

We have already seen how to determine e from the curves of Q and T. These curves are for a fixed diameter and speed of advance and at any given point $P = \dfrac{2\pi RQ}{33000}$.

Now $\qquad R = \dfrac{V \times 101.33}{p(1-s)}$.

Then $\qquad P = \dfrac{2\pi Q}{33000} \times \dfrac{101.33 V}{p(1-s)} = \dfrac{d^2V^3A}{1000}$.

Whence $\qquad A = \dfrac{1000 \times 2\pi Q \times 101.33}{d^2V^2 \times 33000\, p(1-s)}$.

From the experimental results for a model propeller for a given value of s we know everything on the right-hand side of the equation and hence can determine A without difficulty. Similarly, it will be found that we may derive B from the thrust T.

$$E = \dfrac{TV \times 101.33}{33000} = \dfrac{d^2V^3B}{1000},$$

whence $\qquad B = \dfrac{1000\, T \times 101.33}{33000\, d^2 V^2}$.

Then curves of A, B and e completely characterize a propeller. As a matter of fact any one of them can be derived from the other two. They are all functions of slip and proportions and characteristics of the propeller and independent of size and speed. Table XI shows the calculations necessary to determine curves of A, B and e from the experimental data recorded as in Fig. 176.

Fig. 179 shows four curves of A, B and e as deduced from the results of model experiments for four propellers of the same nominal pitch ratio 1.2 and mean width ratio .2 and of the different blade thickness fractions indicated. The curves are plotted upon nominal slip and show that for this blade width and pitch ratio efficiency increases as the blade thickness is reduced, but the power absorption coefficient A and the thrust coefficient B decrease as thickness decreases.

5. Application of Curves of Coefficients from Model Propellers. — Curves of B are particularly valuable in estimating from model results the probable performance of propellers of ships. If there were no reactions between ship and propeller, that is, if the ship were a "phantom ship" as Froude calls it, which offers resistance the same as the actual resistance without disturbing the water or modifying the action of the propeller, the case would be very simple.

For the ship we would know from model experiments the E.H.P. at any speed V and would also know the diameter d of the propeller. Then $B = \frac{1000 \, E}{d^2 V^3}$ is known for any speed from consideration of the ship. But from the propeller model experiments we have a curve of B plotted on slip. So having determined B for a speed V we know what the slip must be. But $R = \frac{101.33 \, V}{p(1-s)}$; hence we know what the revolutions must be. Finally from the slip determined by B we may determine e and A corresponding.

We can then determine the power P absorbed by the screw by either one of the two formulæ. We have

$$P = \frac{E}{e} \quad \text{or} \quad P = \frac{A}{1000} d^2 V^3.$$

We shall see later that the case of the actual ship is not so simple as that of the phantom ship, but curves of revolutions and horse-power deduced entirely from model experiments for phantom ships agree surprisingly well in many instances with the actual curves determined by trial of the full-sized ships.

The author has encountered cases where curves of revolutions and speed obtained from trials of full-sized ships presented features which appeared at first sight abnormal but were found to be duplicated almost exactly by the estimated curves of revolutions and speed deduced entirely from experiments made independently with models of ship and propeller. On the other hand, when the full-sized propeller shows cavitation, the curves deduced from model results differ materially from the actual curves, a fact which in some cases permits of the determination with a good deal of accuracy of the point where cavitation becomes serious.

6. Methods of Plotting Information for Design Work. — The preceding analysis and method of plotting results of model experiments is not very convenient when we come to design work. The designer of a propeller knows in advance or can estimate with reasonable accuracy the power P which the propeller is to absorb and the speed of advance of the propeller through the water V_A. He either knows the revolutions R which are to be used, or, supposing the revolutions may be varied through a certain range, wishes to ascertain the effect of such variation upon his design. He then has to determine diameter, pitch, blade area, blade thickness and blade shape.

It is evident, then, that in plotting model experiments for use in design it would not be advisable to plot them upon slip, because this is not a quantity that is known or can be closely approximated in advance. It is desirable to use variables independent of size but involving power, speed and revolutions, etc. There are many such expressions. For practical applications the following will be found convenient:

$$\rho = R\sqrt{\frac{P}{V_A^5}}, \qquad \delta = d\frac{R^3}{(PV_A)^{\frac{1}{2}}},$$

where d is diameter in feet. The quantity ρ is practically the same as an expression suggested by Mr. R. E. Froude in discussing a paper by Barnaby before the Institution of Civil Engineers,

May 6, 1890 (Vol. CII, p. 101). In discussing the same paper Mr. C. Humphrey Wingfield suggested the use of $\dfrac{R^2 P}{V^5}$.

The quantities ρ and δ may be readily connected with the coefficients already used in analysis of propeller experiments or can be deduced directly from the model propeller results.

Thus we have seen that $P = \dfrac{A}{1000} d^2 V^3$ and we know that $p^2 R^2 (1 - s)^2 = (101\tfrac{1}{3})^2 V^2$.

Multiplying the two together,

$$P p^2 R^2 (1 - s)^2 = \dfrac{(101\tfrac{1}{3})^2}{1000} A d^2 V^5.$$

Whence $\quad \dfrac{R^2 P}{V^5} = \left(\dfrac{d}{p}\right)^2 \dfrac{A}{(1-s)^2} \dfrac{(101\tfrac{1}{3})^2}{1000} = 10.268 \left(\dfrac{d}{p}\right)^2 \dfrac{A}{(1-s)^2}.$

Whence $\quad \rho = R \sqrt{\dfrac{P}{V^5}} = 3.204 \dfrac{d}{p} \dfrac{\sqrt{A}}{1-s}.$

The right-hand expression for ρ is independent of size of propeller, and values of ρ are correctly calculated from a curve of A plotted on s. Usually, however, it is just as convenient to calculate them from the curves of torque, etc., of the model propeller. It will be found that we may write

$$\rho = \dfrac{14.07 \sqrt{Q}}{V_A p^{\frac{3}{2}} (1 - s)^{\frac{3}{2}}}.$$

Similarly, we may write

$$\delta = \dfrac{d R^{\frac{1}{2}}}{(P V_A)^{\frac{1}{2}}} = \dfrac{68.73}{(A)^{\frac{1}{4}}} \left(\dfrac{d}{p(1-s)}\right)^{\frac{3}{2}}$$

$$= \dfrac{41.96 \, d V_A^{\frac{1}{2}}}{Q^{\frac{1}{2}} p^{\frac{3}{2}} (1-s)^{\frac{3}{2}}}.$$

Table XII shows calculations of values of ρ and δ for one of the model propellers whose results are plotted in Fig. 179.

Figure 180 shows for the four propellers of Fig. 179 curves of efficiency and of δ plotted on ρ. The calculations it is seen are made for various values of s, and on the curves of δ the spots corresponding to various values of s are indicated. The scale used for ρ is a variable one, the abscissa values being proportional to $\sqrt{\rho}$ instead of ρ directly. This is a convenient device for spacing widely the values of ρ which are the most important without extending the ρ scale unduly.

The application of curves such as those in Fig. 180 to design work is very simple.

Thus, suppose a propeller is to be designed to absorb 10,000 horse-power with a speed of advance of 20 knots and to have 200 revolutions.

Then $\dfrac{P}{V_A^5} = \dfrac{10000}{3200000} = \dfrac{1}{320}$, $\sqrt{\dfrac{P}{V_A^5}} = \dfrac{1}{17.888}$,

$\rho = \dfrac{200}{17.888} = 11.18$, $d = \dfrac{\delta(PV)^{\frac{1}{4}}}{R^{\frac{1}{2}}} = \dfrac{7.647}{34.2}\delta = .2236\,\delta$.

From Fig. 180 for $\rho = 11.18$ the value of δ for the various blade thickness ratios varies from 54.8 to 57.6, the corresponding values of diameter varying from 12.25 to 12.88.

It is seen, however, that the efficiency is low, only about .66, and the slip high. Evidently the pitch ratio of 1.2 is not adapted to the case and should not be used. But suppose the revolutions desired had been 100. Then we would have

$$\rho = 5.59, \quad d = .3551\,\delta.$$

For this value of ρ we have good efficiency, and if the law of comparison holds we would get good results from a propeller of pitch ratio 1.2. For $\rho = 5.59$ the values of δ range from 54.2 to 58.2 and of d from 19.25 to 20.66. In practice we would choose a value of d corresponding to a blade thickness fraction, then determine the actual blade thickness necessary, and if the resulting blade thickness fraction differed much from that first estimated, a second approximation would be made using the correct blade thickness fractions.

25. Propeller Features Influencing Action and Efficiency

A number of experiments have been made with 16-inch model propellers at the United States Model Basin. Many of the results obtained were published in the Transactions of the Society of Naval Architects and Marine Engineers for 1904, 1905 and 1906.

These results and others not published enable some conclusions to be drawn positively as regards 16-inch propellers and with confidence as regards propellers of ordinary sizes within the limits where the Law of Comparison is applicable.

1. **Number of Blades.** — There were tried a number of propellers with blades identical but differing in number — from two to six. It was found that efficiency was inversely as the number of blades; that is, a propeller with two blades was more efficient than a propeller with three identical blades, that one with three blades was more efficient than one with four identical blades and that one with four blades was more efficient than one with six identical blades.

Also while total thrust and torque increase as number of blades is increased, the thrust and torque per blade fall off. A three-bladed propeller at a given slip does not show 50 per cent more thrust and torque than a two-bladed propeller with identical blades. Fig. 181 shows approximately for working slips the relative efficiencies and coefficients for 2-, 3- and 4-bladed propellers identical except as to the number of blades. The curves are curves of ratios of the quantities concerned, those for 3 blades being taken as unity in each case. As we have seen:

$$A = \frac{1000\, P}{d^2 V^3}, \qquad B = \frac{1000\, E}{d^2 V^3},$$

where d is diameter in feet, V is speed of advance in knots and P and E power absorbed and effective power. The subscripts refer to the number of blades, A_4, for instance, denoting the value of A for 4-bladed propellers. It is seen that the power absorbed, depending upon the coefficient A varies more nearly as the number of blades than the useful horse-power depending upon the coefficient B. The 2-bladed propeller shows slightly greater efficiency than the 3-bladed, and the 4-bladed distinctly less. It should be remembered that Fig. 181 refers to propellers working under identical conditions of slip, speed of advance, etc. This means that a 4-bladed propeller will absorb about 30 per cent more power than a 3-bladed and a 2-bladed propeller about 15 per cent less.

In practice the question to be decided is whether to use a 4-bladed or a 3-bladed propeller when the same power is to be absorbed. In this case the 4-bladed propeller would be smaller than the 3-bladed and hence might have a pitch ratio more favorable to efficiency than the pitch ratio of the corresponding 3-bladed propeller. So the question of 3- or 4-bladed propellers would require investigation in each case. The methods to be used will be considered later.

2. Outline or Shape of Blades. — The question of shape or outline of blade faces has been given much attention in connection with propeller designs and in some cases extravagant claims have been made for special shapes.

Fig. 182 shows five blade shapes which were experimented with at the United States Model Basin. Blade thickness fraction was constant in each case, being .0575. Three pitch ratios were used, .8, 1.0 and 1.2.

The results were quite consistent and showed that the blades with broad tips absorbed more power and gave more thrust but with slightly less efficiency. While the very pointed blades showed up slightly the best, there is some reason to doubt whether they would retain their superiority — which was not very marked — in full-sized propellers. The experiments justify us in looking with doubt upon claims for great gain of efficiency by reason of some special shape of blade, and appear to indicate that for all-round work the old well-known elliptical shape is probably as good as any, though it may be that some other oval shape may be found slightly better. On the other hand the conclusion seems warranted that if circumstances render some special shape desirable, it can be used without serious loss of efficiency provided it is not altogether abnormal.

3. Rake of Blades. — It is a very common practice to rake or incline the blades of a propeller aft. Sometimes they are inclined forward. At the United States Model Basin, six propellers, all of .2 mean width ratio and .0425 blade thickness ratio, were tested. Three were of .6 pitch ratio and three of 1.2 pitch ratio. Of each trio, one had the blades inclined 10° aft, one had the blades set normal to the shaft and one had the blades inclined 10° forward. The diameter was 16 inches in each case. Fig. 183 shows radial sections of the blades. The experiments gave almost identical results, the difference of torque, thrust, and efficiency being slight. So far as efficiency goes, then, there seems no reason to rake the blades of propellers. The advantage sometimes claimed for blades raking aft is that they prevent a supposed centrifugal motion of the water. Careful investigation of 16-inch propellers on test failed to show any evidences of centrifugal action except for some models of very thick blades and coarse pitch tested at 3 knots speed of advance with a slip of 75 or 80 per cent. These models were practically standing still and seemed to throw the water out under the conditions described. Numerous experiments with 16-inch propellers under normal conditions showed the propeller race to be practically cylindrical and that so far from there being centrifugal motion, there is a slight convergence abaft the propeller.

There is little doubt that the advantages of rake as regards prevention of centrifugal motion are imaginary.

A real advantage of rake in practice is that the blade tips of side screws are thereby given greater clearance from hulls of usual form than if the blades were radial or with the same blade clearance strut arms are shorter. A very real disadvantage is the increase of stresses in the blades because of centrifugal action. This will be discussed later. It is a serious matter for quick running screws, and for such screws at least blades should never rake.

4. **Size of Hub.** — One of the features of the Griffith screw introduced some fifty years ago was a large hub — sometimes with diameter a third that of the propeller. These screws were often very successful, and as a result of practical experience there have for many years been advocates of large hubs. Experiments with model propellers at the United States Model Basin have shown that large hubs are distinctly prejudicial to efficiency.

Full-scale experiments with turbine vessels seem to have shown the same thing, material gains in speed having been reported after substituting solid propellers with small hubs for propellers with large hubs and detachable blades. The argument against the large hub is very simple. When a large spherical hub is moving through the water there must be a strong stream line action abreast its center, the water flowing aft. Hence the inner portion of the blades must be working in a negative wake produced by the hub — a condition prejudicial to efficiency.

It is sometimes argued that with a small hub the inner portion of the blades offer more resistance than if they were suppressed and a large hub fitted.

This is probably not true, especially when we consider that the large hub appreciably increases the vessel's resistance. But even if it were true, the prejudicial effect of the large hub upon the blade outside of it would be enough to turn the scale against it.

With slow-running screws of coarse pitch the large hub, while prejudicial to efficiency, will not affect it seriously; but for screws of such fine pitch as usually fitted in turbine work the inner parts of the blades do relatively more work and are relatively more efficient than in the coarse screws. Hence, reduction of the work done by them and of their efficiency through a negative wake set up by a large hub is likely seriously to reduce the efficiency of the screw as a whole.

5. Standard Series of Model Propellers. — We have now considered the minor factors affecting propeller operation and efficiency and will pass to major factors. These are pitch ratio, blade area, blade thickness and slip. In considering resistance of ships the major factors of residuary resistance were investigated by means of a standard series of models whose variations covered the useful range of the major factors concerned. Similarly, the field has been covered for propellers by a standard series of models of varying pitch ratio, mean width ratio, and blade thickness fraction. They were all 3-bladed propellers 16 inches in diameter, with blades that were elliptical in developed outline. The hubs were cylindrical and $3\frac{1}{5}$ inches in diameter, being practically .2 of the propeller diameter. Six pitch ratios were used — namely, .6, .8, 1.0, 1.2, 1.5 and 2.0. For each pitch ratio five blade areas were used. Fig. 184 shows the developed areas of the five blade faces. Their mean width ratios, as shown, were .15, .20, .25, .30 and .35. Six pitch ratios and five mean width ratios resulted in 30 propellers. These were made true screws with ogival blade sections, the backs being circular arcs, and with extra thick blades.

After being tested, the thickness was reduced by taking metal off the back to form new ogival sections, the face being untouched, and thus new propellers with the same faces as before, but thinner blades, were made. These were tested as before. This process was repeated twice, so that each blade was tested in four thicknesses, being finally unusually thin. This made 120 propellers tested in all. Table XIII gives their data. The original propellers are numbered 1 to 30 and the successive cuts denoted by the letters *A*, *B* and *C*. Great care was taken when reducing thickness not to change the face, and toward the edges the recut blades were probably a shade thicker than true ogival sections.

It is difficult to make model propeller experiments with minute accuracy, but in this case, owing to the number of propellers tried and the number of independent variables involved, irregular experimental errors could be practically eliminated by cross fairing on pitch ratio, mean width ratio and blade thickness fraction.

Figures 185 to 208 show the experimental results after this was done in the form of curves of thrust in pounds, torque in pound feet and efficiency. All refer to a 5-knot speed of advance. The results are plotted upon nominal slip as being most convenient for practical applications.

The results of trials of these 120 propellers are worthy of the most careful study. We will now consider them briefly in connection with the influence of pitch ratio, blade area, blade thickness and slip upon thrust, torque and efficiency.

6. Pitch Ratio. — The effect of variation of pitch ratio is illustrated in Fig. 209, which shows for propellers of .25 mean width ratio and .04 blade thickness fraction curves of maximum efficiency and of thrust and torque for 20 per cent slip. This figure is typical. It is seen that for constant slip and speed of advance, torque and thrust increase as pitch ratio decreases, the increase becoming more and more rapid as pitch ratio becomes less.

The efficiency remains nearly constant over a fairly wide range of pitch ratio having its greatest value at a pitch ratio of about 1.5. As pitch ratio decreases, however, efficiency begins to fall off, and below the value of unity the falling off is rapid. In practice screws of fine pitch have frequently shown very low efficiency as a result of cavitation, but apart from this, screws of fine pitch, say below a pitch ratio of unity, are essentially less efficient than screws of pitch ratio 1.5 or so, and the smaller the pitch ratio the less the efficiency.

7. Blade Thickness. — When we study the influence of blade thickness we find that the thicker the blade the greater the thrust and torque for a given slip. This is perfectly natural when we reflect that the results are plotted upon nominal slip and that the thicker the blade the greater the virtual pitch. The effect of blade thickness upon efficiency is summarized in Fig. 210. It was found that for a given blade area the relative variations of efficiency with blade thickness were nearly the same for slips used in practice regardless of pitch ratio. Hence Fig. 210 shows for each blade width an average curve of relative efficiency plotted on thickness only; for each curve, unity corresponds to a different blade thickness fraction, the broad blades being thinner than the narrow blades. This is generally in accordance with what considerations of strength necessitate in practice.

Figure 210 indicates that the efficiency of narrow blades increases rapidly as they are thinned, while for the broad blades thickness has little effect upon efficiency, and in fact the thicker blades seem slightly more efficient. When we remember that on account of strength a narrow blade must be thicker than a broad blade the deduction from Fig. 210 is that practicable variations of blade thickness will have comparatively little effect upon efficiency. This conclusion, however, is from results of experiments where cavitation was not present, and it is generally agreed that to avoid cavitation propeller blades should be as thin as possible.

It is probable that in many cases if the blades are made too thick cavitation would reduce efficiency without the propeller actually breaking down, while it will be avoided altogether with thin blades. Hence we should make propeller blades reasonably thin in practice, in spite of Fig. 210. Where cavitation is likely they must be made thin. It may be remarked, however, that Fig. 210 appears to be in general accordance with facts of experience with slow-running propellers. Coarse, heavy propellers of this type often give very good results in service in spite of thick blades.

8. Blade Areas. — In the experiments with the standard series of propellers it was not practicable to investigate the question of blade area entirely apart from that of blade thickness. The broad blades were made thinner than the narrow ones, as would be the case with actual propellers in practice when it is a question between a narrow-bladed propeller and a broad-bladed propeller to absorb the same power at the same revolutions and speed.

It is owing to the greater thickness of the narrow blades, and hence their greater virtual pitch for a given nominal pitch, that in the fine pitches the narrow blades actually absorb more power and deliver more thrust for a given nominal slip than the broad blades. In the coarse pitches this is not the case for slips such as occur in practice, but the broad blades do very little more than the narrow ones.

Even after making allowances for the thickness effect it is evident that the broad blades by no means absorb torque and deliver thrust in proportion to their areas. In fact the influence of blade area upon thrust and torque is surprisingly small.

Considering efficiency it is seen that for propellers of pitch ratio usually found in practice the broad blades and the narrow blades are both less efficient than blades of medium width, say with a mean width ratio of .25 to .30. The differences are not great, however. It is interesting to note the superior efficiency of the narrow blades for the propellers of abnormally fine pitch. This, however, is not due to the fact that the blades are narrow, but to the fact that the narrow blades have greater virtual pitch ratio, and for the propellers of very fine pitch gain in virtual pitch ratio means gain in efficiency.

The experiments with the standard series of model propellers warrant fully the broad conclusion that, when cavitation is absent, propellers may vary quite widely in pitch ratio (above 1.2 or so) and in area with little change in efficiency, provided diameter is such that they work at slips at or near that of maximum efficiency. This conclusion is fully borne out by experience, which has led many people to conclude that there was so little difference between propellers that any propeller which allowed the engine to develop its power at the desired revolutions and showed a good slip was a good enough propeller. For low-speed work this is reasonably correct; for high-speed work, even leaving out of question cavitation, propellers which absorb the power at the desired revolutions are liable to vary seriously in efficiency, particularly if, as is usually the case, they must be of the fine pitch type.

9. **Slip.** — Figures 185 to 208 show that all curves of efficiency plotted upon slip present the same general appearance. Considering nominal slip the efficiency is zero at a certain negative slip. The thicker and narrower the blade the greater in general the increase of virtual over nominal pitch, and the greater the numerical value of the negative slip corresponding to zero efficiency. It will be noted, however, that for the narrow blades of pitch above unity there seems to be a slight falling off of virtual pitch with thickness beyond the A cut. This is probably due to the fact that as the thickness of these narrow blades is increased a point is reached where the water breaks away from the back, the latter losing its grip, as it were. The process is analogous to cavitation, though cavities are not formed. As the slip increases from that corresponding to zero efficiency, the efficiency rises very rapidly at first, then reaches a maximum and thereafter falls off. The nominal slip corresponding to maximum efficiency is nearly always between 15 and 20 per cent for blade thickness that would be used in practice, but slip can be increased to 25 per cent, and in some cases to 30 per cent, without serious loss of efficiency.

But such an increase means an eno mous increase in thrust and torque. Hence a given propeller will vary widely its power and thrust without material change of efficiency. So it is not necessary in practice with propellers of coarse pitch, to aim very closely at some exact slip provided the propeller is so designed that under conditions of service its slip is not too small. A propeller which is too large, showing slip much below that for maximum efficiency, will be very inefficient. On the other hand, a propeller may be too small and work with slip a good deal greater than for maximum efficiency without much loss of efficiency. It should be remembered that the slips of Figs. 185 to 208 refer to propellers operating in undisturbed water, and the apparent slip of propellers attached to ships is usually less than the true slip.

When dealing with propellers of fine pitch ratio, say in the neighborhood of unity, the question of efficiency as affected by slip is complicated by the question of efficiency as affected by pitch ratio. Thus in Fig. 190 we see that propeller No. 8, A cut, of .25 mean width ratio and .8 pitch ratio has a maximum efficiency of .632 at 15 per cent slip. From Fig. 194, propeller No. 13, A cut, of .25 mean width ratio and 1.0 pitch ratio has a maximum efficiency of .684 at 14 per cent slip and an efficiency of .632 at about 31 per cent slip. In a given case, then, where we could fit a propeller of the proportions of No. 8 working at maximum efficiency, we could make an improvement if we could fit a propeller of the proportions of No. 13 working below its maximum efficiency provided its slip did not exceed 30 per cent. This is a question of very considerable practical importance. In the next section will be given methods for determining the best combinations of pitch ratio and slip for given conditions.

26. Practical Coefficients and Constants for Full-sized Propellers Derived from Model Experiments.

1. General Line to be Followed in Reducing Model Results. — The results of the model experiments for the standard elliptical 3-bladed series will of course be of value in the case of any propeller design. It should be carefully remembered, however, that they cannot be applied blindly. We have determined experimentally the thrust and torque and deduced the efficiency of a number of small propellers at a 5-knot speed of advance throughout the range of slip likely to be found in practice. These small propellers covered for 3-bladed elliptical propellers the range of pitch ratio, mean width ratio, and blade thickness fraction likely to be found in practice. We know that so long as cavitation does not appear the Law of Comparison will apply satisfactorily and that the results of the model experiments will apply to full-sized propellers working under the same conditions as the models. But in applying the results we must remember that they do not hold for cavitating conditions, which will presently be considered separately.

The models were tested in such a manner as to be practically free from hull influence, and we know that for full-sized propellers driving ships there are material mutual reactions between propeller and ship. The question arises whether we shall attempt to take account of these reactions in reducing the model results or consider them separately.

It is much better, and even simpler in the end, to attack the problem in detail.

2. Reduction of Model Results. — We have seen that by means of a $\rho\delta$ diagram, as in Fig. 180, the experimental model results may be reduced to a form convenient for practical applications. But if we simply construct a $\rho\delta$ diagram for each model tested it will be a very laborious process to locate and utilize the particular diagram adapted to a particular case. So it is necessary to develop diagrams, by interpolation if necessary, such that the primary factors involved are readily determined. We have to deal with efficiency, diameter, pitch ratio, mean width ratio and blade thickness fraction.

These are too many variables to be covered directly on a single diagram. The first three are the most important. Width and blade thickness are not independent in practice. To do a given work at given revolutions the narrow blade must be thicker than the wide blade. So four $\rho\delta$ diagrams, Figs. 211 to 214, have been constructed from the model results of Figs. 185 to 208. Figure 211 refers to blades having a mean width ratio of .20 and a blade thickness fraction of .06. Similarly Figs. 212, 213 and 214 refer respectively to mean width ratios of .25, .30 and .35 and blade thickness fractions of .05, .04 and .03. We shall see later how to make slight changes involved by other blade thickness fractions.

The application of the $\rho\delta$ diagrams is very simple:

$$\rho = R\sqrt{\frac{P}{V_A{}^5}}, \qquad \delta = d\frac{R^3}{(PV_A)^{\frac{1}{2}}},$$

where P is the power absorbed by the propeller of diameter d feet at R revolutions per minute when advancing at a speed of V_A knots.

Then ρ is the primary variable fixed by the conditions of the problem. Contours of δ are plotted above ρ for equal intervals of pitch ratio and curves of efficiency for the same intervals. When ρ is known we can determine very promptly for any value of δ the pitch ratio and efficiency. In addition to the contours of δ above ρ contours of slip are plotted in dotted lines.

3. **Maximum Efficiency.** — The efficiency curves show many interesting and significant features. For a short interval each pitch ratio shows an efficiency greater than any other, and evidently if our choice is free we should for a given value of ρ use the pitch ratio corresponding to optimum efficiency. Hence, there is drawn an enveloping curve of maximum efficiency touching the successive efficiency lines for the various pitch ratios which has upon it a scale of the pitch ratios for maximum efficiency.

In this connection attention may be called to the fact that the portion of each efficiency curve which gives the best efficiency for a given ρ is in general of an efficiency below the maximum efficiency attainable with the pitch ratio. This is particularly noticeable for the largest values of ρ. For all values of ρ above very small ones it is better to use a propeller of relatively coarse pitch and work it at a fairly high slip — greater than that corresponding to its maximum efficiency — than to use a propeller of finer pitch and work it at its maximum efficiency. This for the reason that for propellers of pitch usual in practice decrease of pitch means falling off in efficiency.

The $\rho\delta$ diagrams bring out clearly some of the basic conditions affecting propeller design.

Once we fix for a propeller the power, P, it is to absorb, its revolutions per minute, R, and its speed of advance, V_A, the value of ρ is fixed. Now it is apparent from the diagrams that for a given value of ρ there is a maximum efficiency beyond which we cannot go. We may very easily fall short of it, but even if we adopt the very best combination of diameter, pitch and blade area possible, we cannot get beyond a limiting efficiency. The $\rho\delta$ diagrams of Figs. 211 to 214 were deduced from experiments with models of 3-bladed propellers with elliptical blades having ogival sections. Hence the limiting efficiencies shown in them are not exactly the same as for all types of propellers, though they are about as high as for any known type. But there is no doubt that they indicate well the general variation of efficiency with ρ for all types of propellers in present use. While there is a maximum efficiency, about $\rho = 3$, and the efficiency falls off on either side, the values of ρ that are found in practice are almost never materially below 3, so that in practice the larger the ρ the smaller the limiting efficiency. It is the high value of ρ produced, if we give low-speed vessels high revolutions, that has hitherto prevented the application to cargo vessels of turbines directly connected to the propeller. Thus, suppose we had a destroyer propeller absorbing 5000 shaft horse-power at 800 revolutions with a speed of advance of 30 knots. For this case

$$\rho = \frac{800 \sqrt{5000}}{(30)^{2\frac{1}{2}}} = 11.5.$$

The limiting efficiency for this value of ρ is about .65 which though low is not impossible. If now we had a large single-screw cargo and passenger vessel which required 5000 shaft horse-power to make 15 knots speed of advance and adhered to 800 revolutions per minute the value of ρ would be

$$\frac{800 \sqrt{5000}}{(15)^{2\frac{1}{2}}} = 64.9.$$

For this value of ρ the limiting efficiency would be inadmissibly low. To hold ρ at 11.5 the revolutions would have to be reduced to 142 which would make an inefficient turbine. An alternative is to hold revolutions at 800 and use multiple shafts. But in order to make the ρ value for each propeller 11.5 only, it would be necessary to divide the 5000 shaft horse-power between 32 shafts, which is of course impossible.

Another fact of serious practical importance which the $\rho\delta$ diagrams bring out is that there is practically a lower limit to the pitch ratio which can be used to advantage. At first the best pitch ratio falls off rapidly with increase of ρ, but for large values of ρ the pitch ratio falls off more and more slowly, and for no value of ρ which it would be advisable to use in practice is it desirable to go below a pitch ratio of .9 or a little less.

The slip for the best all-round efficiency which is below .15 for small values of ρ increases steadily, until it is seen that propellers of a pitch ratio of .9 should be worked at over .30 slip. This is real slip, not apparent slip.

It is interesting to note in this connection that the model experiments indicate that the broader the blades the greater the slip for the best results. Thus for a pitch ratio of 1.0 and the four blade width ratios of .20, .25, .30 and .35 the best slips are respectively .255, .265, .280 and .320. This is in accord with theoretical considerations.

4. **Methods of Calculations.** — In order to facilitate the calculation of ρ in a given case there are given in Table XIV values of $V_A^{2.5}$.

It should be carefully borne in mind that V_A is not the speed of the ship through the water but the speed of advance of the propeller through the disturbed water in which it works. The difference between V_A and V, the speed of the ship, will be considered in connection with the wake factor.

The formula for δ is

$$\delta = d \frac{R^{\frac{2}{3}}}{(PV_A)^{\frac{1}{2}}},$$

or, when δ has been determined,

$$d = \delta \frac{(PV_A)^{\frac{1}{2}}}{R^{\frac{2}{3}}}.$$

With a table of squares and cubes we can readily determine $(PV_A)^{\frac{1}{2}}$ by taking the square root of the cube root of PV_A; $R^{\frac{2}{3}}$ is simply the square of the cube root of R. Hence the calculations required in connection with the use of the $\rho\delta$ diagrams are readily made.

5. **Blade Thickness Correction.** — The four $\rho\delta$ diagrams for the standard series refer to a definite blade thickness fraction for each mean width ratio. We have seen in Fig. 210 the effect upon the efficiency of the standard series of variations of the blade thickness. This effect is not large enough to be of practical importance in most cases. But variation of blade thickness will also necessarily affect pitch ratio and diameter. Investigation shows, however, that the effect is not large, and for blade width ratios from .25 to .35, and for propellers of about the proportions for maximum efficiency, the average corrections required are shown in

Fig. 215. The curves of this figure give for various values of ρ the percentages by which diameters and pitches determined from the $\rho\delta$ diagrams must be modified when the standard blade thickness fractions to which the $\rho\delta$ diagrams correspond are departed from.

The corrections are small and in practice may often be ignored. The standard $\rho\delta$ diagrams already take some account of thickness, the widest blades being only half as thick as the narrowest, but of course the actual blade thickness fraction in a given case is fixed mainly by considerations of strength.

6. **Four-bladed Propellers.** — The standard $\rho\delta$ diagrams, Figs. 211 to 214, refer to three-bladed propellers. It would be desirable to have similar diagrams from full experiments with four-bladed propellers, but lacking such they can be used with fair approximation for four-bladed propellers. We have in Fig. 181 the relation between power absorbed, thrust and efficiency of three and four-bladed propellers as deduced by analysis of experiments at the model basin with propellers having quite thin blades of rather broad tips. These may be taken as applying with reasonable approximation to the elliptical blades.

Then the steps in a given case will be as follows:

1. Determine ρ in the ordinary way and then divide it by the square root of the ratio between the coefficient A for a four-bladed screw and for a three-bladed screw — these ratios are given in Fig. 181. Call the quotient ρ_4.

2. Using ρ_4, determine by the use of the proper $\rho\delta$ diagram the proper diameter, pitch, etc., for a three-bladed propeller.

Then upon adding a fourth identical blade to the three-bladed propeller we shall have a four-bladed propeller which will meet the conditions.

For let P, R, and V_A denote power to be absorbed, revolutions to be made and speed of advance.

We have
$$\rho = R \frac{\sqrt{P}}{V_A^{2.5}},$$

then
$$\rho_4 = \frac{R\sqrt{\dfrac{P}{r}}}{V_A^{2.5}},$$

where r is the ratio of the A coefficients from Fig. 181. Then a three-bladed propeller based upon ρ_4 will, at revolutions R and speed of advance V_A, absorb a power $\dfrac{P}{r}$. But from Fig. 181 again a four-bladed propeller identical as to diameter, pitch and blades will absorb r times the power of the three-bladed one, or $\dfrac{P}{r} \times r = P$. Hence the four-bladed propeller will absorb the power P at revolutions R and speed of advance V_A. The relative efficiencies may be obtained from Fig. 181.

Since once we know ρ, we can determine the relative diameters of the three and four-bladed propellers; we can from each $\rho\delta$ diagram for three-bladed propellers determine, by using Fig. 181 as explained above, a figure giving ratios of diameter, pitch and efficiency for three and four-bladed propellers. It is found, however, that as regards diameter and pitch the ratios are so nearly the same for all widths that the results may be averaged in a simple diagram (Fig. 216).

This gives curves of coefficients by which the diameter and pitch of a three-bladed propeller must be multiplied to determine the diameter and pitch of a four-bladed propeller of the same type of blades and mean width ratio that at the same revolutions and speed of advance will absorb the same power.

Efficiency coefficients are also given. These are seen to be all less than unity, indicating a loss of efficiency by adopting four-bladed instead of three-bladed screws.

The pitch coefficient is less than unity throughout, so the pitch of the four-bladed screw will be slightly less than that of the three-bladed screw, but the diameter is reduced more than the pitch, so that the pitch ratio of the four-bladed screw will be the greater. The diameter coefficient in Fig. 216 should be regarded as an upper limit. It will be feasible in practice to reduce the diameter of the four-bladed screw four or five per cent more without material loss of efficiency.

7. Two-bladed Propellers. — It is evident that the methods above may be utilized in order to apply the $\rho\delta$ diagrams for the three-bladed propellers to two-bladed propellers.

In this case, however, the artificial value of ρ will be greater than the original value.

Fig. 217 gives curves of coefficients, etc. It is seen that diameter, pitch and efficiency are all increased. The gain in efficiency is small, however, and there are practical objections to two-bladed propellers, so that their use is seldom expedient. This point will be discussed further in considering design of propellers.

27. Cavitation

1. **Nature of Cavitation.** — The phenomenon known as cavitation has been given much attention of late years in connection with quick-running turbine-driven propellers. It appears to have been first identified upon the trials in 1894 of the torpedo boat destroyer *Daring* which had reciprocating engines. When driven at full power with the original screws this vessel showed very serious vibration evidently due to some irregular screw action. The propulsive efficiency was poor, the maximum speed obtained being 24 knots for 3700 I.H.P. and 384 revolutions per minute.

Mr. Sidney W. Barnaby, the engineer of the *Thorneycrofts*, who built the *Daring*, came to the conclusion that at the high thrust per square inch at which the screws were working the water was unable to follow up the screw blades and that " the bad performance of the screws was due to the formation of cavities in the water forward of the screw, which cavities would probably be filled with air and water vapor." So Mr. Barnaby gave the phenomenon the name of cavitation. The screws which gave the poor results had 6 feet 2 inches diameter, 8 feet $7\frac{3}{4}$ inches pitch and 8.9 square feet blade area. Various alternative screws were tried, and the trouble was cured by the use of screws of 6 feet 2 inches diameter, 8 feet 11 inches pitch and 12.9 square feet blade area. With these screws 24 knots was attained with 3050 I.H.P. and the maximum speed rose from 24 knots to over 29 knots.

For the *Daring* cavitation appeared to begin when the screw area was such that the thrust per square inch of projected area was a little over 11 pounds per square inch. For a time it was thought that the thrust per square inch of projected area was a satisfactory criterion in connection with cavitation and that the limiting thrust per square inch of projected area found on the *Daring* was generally applicable.

This, however, is not the case. Greater thrusts have been successfully used and cavitation is liable to appear at much lower thrusts. In one case within the author's experience cavitation appeared when the thrust was about 5 pounds per square inch of projected area, the tip speed being about 5000 feet per minute, and in another when it was about 7.5 pounds, the tip speed being about 6500 feet per minute. There is little doubt that the prime factors involved in cavitation are: (1) the speed of the blade through the water, which is conveniently measured by the tip speed, and (2) the shape of the blade section.

2. Accepted Theory of Cavitation Inadequate. — When we attempt to explain just how or why vacuous cavities at the backs of screw blades cause the serious loss of efficiency associated with cavitation we encounter insuperable difficulties. Suppose, for instance, the cavity is a vacuum and covers the whole blade back. Then the thrust per square inch of projected area due to the vacuum on the blade back would be between 14 and 15 pounds and the thrust due to the face would be added to that. As cavitation will appear in some cases at thrusts per square inch of projected area as low as 4 pounds, it is evident that in such cases there cannot be a vacuum over the whole blade back and thrust in addition on the face.

But suppose the blade had a vacuum over a portion of the back only. There would be no increase of thrust from additional suction of that portion of the blade back, but neither would there be any increase of torque due to that portion of the blade back. The only loss of efficiency would be a small amount due to the propeller working with a slightly higher slip, while the loss of efficiency accompanying cavitation is very much greater than this.

Fig. 218 shows a propeller blade section advancing through the water at an angle of slip of 3 degrees — not an unusual angle. There are three regions indicated:

1. The leading portion of the back, denoted by A.
2. The following portion of the back, denoted by B.
3. The face, denoted by C.

It does not appear possible that cavities can form at A. This portion of the section contributes negative thrust, and although the point of demarcation between the portion of the back contributing negative thrust and the portion contributing positive thrust (suction) probably varies in position with speed through the water and slip angle, it appears reasonably certain that A always contributes negative thrust and quite probable that this negative thrust increases indefinitely with the speed.

Over B a cavity will form when the speed is high enough. It will probably be small at first, and as the speed is increased, cover a greater and greater portion of the section back. It cannot cover the whole back, however, because it cannot extend over A to the leading edge.

As regards C it has been generally assumed that the thrust from the face always increases with increase of speed of the section through the water.

3. **Possible Theories of Cavitation.** — Now how is it conceivable that cavitation can cause a rather sudden loss of efficiency when the section is pushed to a sufficiently high speed?

A. It is possible that when a vacuum is formed at B this portion of the blade contributes no more suction or thrust while the negative thrust at A continues to increase with resulting loss of efficiency. This explanation would seem to involve the further assumption that by far the major portion of the thrust of a propeller is due to the suction of the blade back.

B. It is possible that when a vacuum is formed at B it is spoiled by air obtained from the surrounding water and the suction of the blade back is decreased. This explanation is possible only if, when the water still hugs the blade back, it sweeps away any air which is sucked out of the water, so that while the water is in contact with the back it is possible for the latter to exert a suction approaching that of a perfect vacuum. But when the water breaks away from the back, air leaking into the space is carried away by entrainment only from the rear of the cavity, where the water comes together again; and when the rate of entrainment is equal to the rate of leaking into the cavity there is a balance of pressure, and though there is a partial vacuum in the cavity the pressure is much greater than a complete vacuum.

C. It is possible that when cavitation sets in the thrust from the blade face falls off absolutely or relatively.

A, B and C above appear to cover the possible theories of the phenomena associated with cavitation. Whether cavitation is due to one or more of these explanations or to something different still, can be satisfactorily determined by experiment only, either on models or on full-sized propellers.

4. **Experimental Investigation of Cavitation.** — Experiments with cavitation using full-sized propellers have not hitherto been made, except inadvertently. While no theory of cavitation should be fully accepted until confirmed by full-sized experiments the expense of a general investigation with large propellers has been hitherto prohibitive, to say nothing of the time required and the practical difficulties in the way. Small scale or model experiments on cavitation present special difficulties. For the law of comparison to apply in spite of cavitation it would be necessary to have the pressure around the model in the ratio of the size to the pressure around the full-sized propeller.

This requires the model to work in water whose surface is covered by a partial vacuum, or in hot water which has a vapor pressure partially neutralizing that of the air.

The Hon. C. A. Parsons has done some work using the latter method, but little has been published of the results. There are great practical difficulties in making experiments along this line, except with very small models.

A second possible method of investigating cavitation experimentally by means of models is to test the model, not at the corresponding speed, but at the actual speed of advance of the full-sized propeller. When this is done, the pressures per square inch at corresponding points of propeller and model are the same, and if one shows cavitation so will the other. This method is hardly practicable for the model of the propeller of a 33-knot destroyer, but for propellers of slow and moderate-speed vessels experiments could be made without serious difficulty or great expense, either in a model basin or from a special testing platform in front of a vessel. This method, however, has not been used in practice. For model propellers of any size, say 15 inches to 18 inches in diameter, it would require very powerful driving and measuring gear.

A third method is to use the propeller testing gear already installed in a model basin with small propellers of such abnormal proportions and shape that they will show cavitation within the limits of speed and revolutions available.

Some experiments along this line have been made at the United States Model Basin.

To obtain pronounced cavitation from small propellers 12 inches to 16 inches in diameter, tested at speeds of advance not over 7 knots or so, it is necessary to make the pitch ratio much smaller and the ratio of thickness to width of blade much larger than for the propellers used in practice. Sixteen-inch models representing propellers of ordinary proportions will not cavitate satisfactorily at low speeds of advance, and the experimental gear available was not powerful enough to drive them at high speeds.

The results obtained with the fine pitch propellers appear, however, to throw some light upon the subject under consideration.

Figure 219 shows expanded blade outline and blade sections for a 16-inch model propeller of 6.4-inch pitch. Figure 220 shows curves of thrust and torque for this propeller plotted upon slip for speeds of advance of 5, 6 and 7 knots. The major portion of Fig. 220 is from Fig. 10 of a paper by the author before the Society of Naval Architects and Marine Engineers in 1904, but the curves for the 5-knot speed have been extended, and the curves for the propeller reversed have been added from the results of subsequent experiments. For the propeller reversed the nominal slip is figured from the nominal pitch of the back as tested (the face before reversal).

Figure 220 shows conclusively that, so far as this propeller is concerned, the thrust per square inch of projected area has little to do with the cavitating point. At a nominal slip of −15 per cent there is evidently cavitation at the 7-knot speed. At this point the thrust is 80 pounds, or almost 4.3 pounds per square inch of projected area. At 5 knots, however, the thrust per square inch of projected area at which cavitation begins is about 9 pounds.

Other conclusions might be drawn from Fig. 220, but more illumination can be obtained from the results of trials of a small propeller especially designed to show cavitation. This propeller was 14 inches in diameter and of 4.2 inches pitch. Its developed blade outline and blade sections are shown in Fig. 221. At the points A, B, C, D and E small holes were made on each blade connecting to the shaft, which was hollow. The hole in the shaft communicated in turn with a pipe forward of the hub, which led finally to a tank under air pressure, there being a pressure gauge on the line and valves for turning on or cutting off the air pressure as desired. When making trials one hole only was left open in each blade. This apparatus measured suction or partial vacua with great facility but had to be handled carefully to measure pressure. When measuring suction, the air pressure was cut off, when the propeller itself would quickly exhaust the air and the amount of vacuum was read on the gauge. When measuring pressure, the air valve was barely cracked, so that a small quantity of air was dribbling out all the time through the hole where pressure was to be measured.

In this way the passages in the propeller were kept clear of water, whose presence would have prevented obtaining the pressure at the hole.

A gauge pressure of a pound and a half or so was sufficient to keep the air passing out when the propeller was at rest or turning over very slowly, and the difference between this initial pressure and the gauge pressure shown while running was taken as pressure at the hole.

In the early part of a run for pressure the air would stop coming out of the propeller; it would accumulate in the pipe and the gauge pressure rise until air again began to come out and the gauge became steady. At the end of a run the instant the propeller began to slow down the air would burst forth.

While the apparatus and methods described above for measuring pressure and suction are certainly not of minute accuracy, they gave consistent results which are believed to be reasonably accurate.

For looking at the propeller under the test there was fitted a fixed disc with a small slot, and immediately behind it a revolving disc with a similar slot, which was driven at the same speed as the propeller. The propeller was illuminated by a searchlight and when looking through the slot in the fixed disc the propeller was seen once during each revolution always in the same position. The discs and searchlight could be shifted so that either back or face of the propeller could be observed.

Figure 222 shows for the propeller of Fig. 221 and three knots speed of advance curves of thrust, torque and of pressure or suction at the points indicated. The curves are plotted upon nominal slip and pressure and suction are measured in pounds per square inch. A scale showing tip speed is also given.

Figure 223 gives the same data as Fig. 222 for five knots speed of advance.

When watching the operation through the slotted discs any cavities present were plainly visible and it was easy to trace the development of cavitation.

At about 3000 feet tip speed cavities appeared at the following portions of the back and the leading portions of the face. The cavities appeared first on the face, as might be expected from Figs. 222 and 223, which show that the suction at A is always more intense than at D.

The cavities first show themselves near the blade tips and creep in toward the center as speed is increased.

In Figs. 222 and 223 the thrust has returned to zero, when the tip speed is between 5000 and 6000 feet per minute. When this is the case the cavities at the back of the blade extend in from the tip about two-thirds of the blade length and near the tip cover nearly two-thirds of the blade back.

On the face under the same conditions the cavities extend along the leading edge practically in to the hub and near the tip from the leading edge to the following edge.

5. **Theory and Cause of Cavitation.** — From the experimental curves of Figs. 222 and 223 and observation of the cavities it is obvious that the cavities at the rear of the blade do no harm. It is the cavities on the driving face which grow rapidly as tip speed is increased, combined with the negative thrust of the leading portions of the blade back that stop the increase of thrust and then actually cause it to decrease to zero and below.

These conclusions apply strictly to the 14-inch model propeller of somewhat abnormal type shown in Fig. 221, but it seems reasonably certain that they apply more generally, and that harmful cavitation is due not to cavities at the backs of propeller blades, but to cavities at their driving faces.

When we seek a cause for these cavities, it seems fairly obvious. Fig. 218 shows the section of a propeller blade advancing with a slip angle of 3 degrees, which is not an exceptionally small angle, as is evident from Fig. 170. But the face C, advancing through the water at an angle of 3 degrees, is associated with the leading portions of the back, whose direction is such that it is advancing through the water at an angle of over 20 degrees. Fig. 63 shows diagrammatically the nature of the motion of water past a plane with a sharp edge. In the case of the propeller we have virtually two planes in association; namely, the face and the leading portions of the back. Considering the face alone, the water tends to cascade around the leading edge from front to back. Considering the back alone, the water tends to cascade around the leading edge from back to front. Acting in association, the back of the blade with an inclination of over 20 degrees overpowers the face with an inclination of 3 degrees, and as a result the water cascades from the back of the blade to the face around the leading edge, causing first eddies and then cavities on the face of the blade.

In regarding the leading portion of the propeller blade as made up of two planes, we should remember that the motion at each point is circular, not linear. A plane in linear motion can drag a good deal of dead water behind it, and water brought to rest relatively to the plane passes aft again without any motion across the plane. The propeller blade is moving in a circle and cannot carry water with it in the shape of " dead " water for any distance. Centrifugal action would rapidly throw it out, and no doubt strong centrifugal force acts upon the water which is brought nearly or entirely to rest relatively to the blade by impinging upon the leading edge. It is possible that this strongly localized centrifugal force plays a part in causing cavitation.

It is evidently necessary to consider separately the cavitation which appears over the backs of propeller blades and the cavitation which appears over the faces.

The former is not seriously objectionable. If the cavities at the blade backs were perfect vacua they would be helpful rather than harmful. It is seen from Figs. 222 and 223 that for model propellers in the fresh water of the model basin these cavities do approach perfect vacua. Sea water contains a good deal of occluded air, and it may be that for full-sized propellers in sea water the cavities are more or less filled with air. But, even so, the air could be pumped out without serious difficulty. Hence we may conclude that cavities at the rear of a blade are not an insuperable bar to efficiency. This is fortunate, for there is no question that when a curved surface, such as the back of a propeller blade, is driven through the water at a sufficiently high speed, cavities are necessarily formed over its rear portions.

The case of the cavities over the blade faces is different. These have no redeeming feature. In the first place, they are due to an edge angle so large as to produce large negative thrust from the leading portion of the back of the blade. In the second place, they nullify the thrust which the blade face would otherwise contribute, and, all things considered, are obviously fatal to efficiency.

Hence, it is essential to efficiency to minimize or avoid entirely face cavitation. The method which has been most used with satisfaction in practice consists in fitting very broad blades so that the thrust per square inch of projected area is kept below a limit found to be safe by experience. But the thrust per square inch of projected area is not the primary feature causing cavitation. Tip speed and blade section are without doubt the main factors. Still, for a given type of propeller the thrust is a function of tip speed and blade section, and hence might be used as a gauge of cavitating conditions. Thus Barnaby, for the type of propeller used on the *Daring*, found that with a tip immersion of one foot, cavitation showed up when the thrust per square inch of projected area was above $11\frac{1}{4}$ pounds. The trouble with this method is that the limiting thrust permissible would have to be determined for each type of propeller.

6. **Reduction of Cavitation by Broad Blades.** — From the theory of cavitation set forth above the advantages of a wide, thin blade are obvious. It has a smaller edge angle, so that it can be driven to a much higher tip speed than a narrow blade without causing

face cavitation. Also after face cavitation begins it spreads slowly with increase of tip speed so that the wider the blade the greater the area of the face whose thrust is not nullified by cavitation.

In fact, if the blade is so wide that the manner of the water leaving it is not materially modified by cavitation, the thrust will not be materially modified even if there is a cavity over the leading portion of the face. This result is readily explicable. Thus, suppose we have a cavity at the leading portion of a blade face. The vacuum results in the water being impelled toward the face, forward momentum being communicated to it. If the face is sufficiently wide, the water will impinge upon it again. Through the loss of its momentum it will communicate a corresponding thrust to the blade, and then will pass from the blade, if it is wide enough, in nearly the same manner as if there were no cavitation over the forward portion of the face. Hence, the net change of velocity and resulting thrust will not be much affected by the cavitation. But if the blade is so narrow that the face cavity extends nearly to the following edge there will not be enough blade beyond the cavity to absorb the forward momentum of the water and direct it again in the way it should go. With the wide blade the loss of pressure on the leading portion of the face due to cavitation is nearly made up by additional pressure on the following portion of the face. With the narrow blade there is virtually no following portion.

Figures 224 and 225 show experimental results which indicate the advantages of breadth of blade in preventing harmful effects from cavitation. Two 16-inch model propellers of the same pitch ratio — 0.4 — and blade thickness fraction, but of mean width ratios of .125 and .275, were tested with smooth backs and with strips secured to the backs, as indicated in the figures. The sections shown were taken in each case at two-thirds the radius. The curves in each case refer to a 5-knot speed of advance. Neither propeller showed harmful effects of cavitation with a smooth back. With the strip attached the narrow-bladed propeller showed pronounced cavitation, while the broad-bladed propeller showed none, though its strip was materially larger than that of the narrow-bladed propeller. As might be expected, the torque is much increased by the presence of the strip. But until cavitation appears the thrust of the narrow-bladed propeller is but little reduced by the strip, and for the broad-bladed propeller the thrust is actually increased by the presence of the strip. Upon the theory of cavitation which has been set forth a reasonable explanation of the peculiar features of Figs. 224 and 225 is as follows:

The strips increase the negative thrust on the leading portion of the blade back in each case, increase the suction or cavitation of the following portion of the blade back, thus increasing thrust, and cause face cavitation over the leading portion of the blade face. The net result of the two former actions is small or even results in an increased thrust. But when face cavitation is set up strongly, the narrow blade breaks down, while the broad blade holds its own, because the face cavitation over the leading portion of the face is neutralized by the action of the following portion of the face.

7. **Cure for Cavitation.** — We have seen that the wide blade of usual type has two advantages from the point of view of cavitation. Its smaller edge angle will allow high tip speeds to be reached without cavitation, and when cavities do appear the tip speed can be still further increased without the harmful effects due to the face cavities, which are usually characterized by the term "cavitation." Now we do not mind cavities on the back of the blade, so the question whether it is possible fully to cure harmful cavitation depends entirely upon whether it is possible to avoid entirely face cavitation.

The difficulties in the way of this are practical difficulties of construction. Thus, if we could make propeller blades without thickness, there would be no face cavitation. The water would cascade around the leading edge from front to back. There would be back cavitation only, and solid water over the face. But we cannot make propeller blades of no thickness. The best we can do in practice is to approximate to the ideal plane along the leading edge, making the face straight, or very slightly convex, and the leading portions of the back hollow, as indicated in Fig. 226, and keeping the edge angle down as close as possible to the slip angle.

It might seem that the edge angle could be made double the slip angle without danger of face cavitation, since when so made the edge would part the water evenly. But the slip angle is an average angle, and usually at some part of its revolution the blade of an actual propeller will have a slip angle but little if any greater than half the average value. Another reason for making the edge angle as small as practicable is the fact that no matter how sharp the edge is made it is not a mathematical edge, and when advancing at enormous speed through the water will show slight cavitation if it is attempted to split the water evenly on each side. Hence, the endeavor should be to have the water naturally tend to cascade around the edge from face to back.

It might seem that this could be accomplished without extreme sharpening of the leading edge by making the leading portion of the face convex, as indicated in Fig. 227. This is true, and a propeller so shaped would not show face cavitation near the leading edge, but with even a moderate convexity of the face it would show severe cavitation over the following portion of the face. There was a case of a United States battleship whose propeller did not differ materially in dimensions, etc., from those of her sister vessels, but had sections which were abnormally curved at the leading portion of the face, as indicated in Fig. 228.

This vessel showed over a knot less speed than her sister vessels for the same power, and although her tip speed was only about 6000 feet per minute, there is little question that she showed very serious face cavitation. It is not possible to say what convexity is permissible in a given case without cavitation, but it is certain that the higher the tip speed the smaller the permissible convexity, and for tip speeds of 10,000 feet and over it probably should be very small indeed. Pending careful full-scale experiments on this point, the safest plan is to avoid convex blade faces for propellers of high tip speed.

It need hardly be said that it is not easy to make hollow-backed propellers with leading edges as sharp as a knife. It is advisable to use cylindrical ribs on the back, extending from the leading edge to the thicker portion of the blade. If the leading edge is serrated with a rib extending to the point of each tooth, the blade edge need not be quite so sharp. Such a form of edge seems to get through the water with less tendency to face cavitation, and when this does set in it seems to confine itself to rather narrow rings, starting from the angles where the roots of the serrations join.

The ribs on the back must of course be well sharpened where they cut the water. They increase back cavitation, but that is not a very serious matter.

While the prevention of face cavitation is essentially a question of the extreme leading portion of the blade back, the blade should not thicken so rapidly as we pass aft from the hollow portion that owing to its angle of action there is large negative thrust.

This is of course always objectionable, but particularly so when there is pronounced back cavitation. After this has set up, the suction of the back does not grow so rapidly as before with increase of speed, and hence negative thrust, which continues to increase indefinitely with speed, should be avoided with peculiar care.

The practical conclusion in this connection is that blades made hollow-backed to avoid cavitation should not be of narrow type but fairly wide — say from .30 to .35 mean width ratio — in order that they may be made fairly thin in the center.

Such blades should avoid cavitation without the excessive widths which are necessary with blades of ogival section and which involve material loss of efficiency through large blade friction.

8. Pressure Due to Blade Edge Speed. — In connection with the question of cavitation it is interesting to note that at the tip velocities of modern high speed propellers enormous pressures are liable to be set up upon the leading blade edges. Suppose we have a small plane advancing through water perpendicular to itself. The maximum pressure upon it is that due to a head equivalent to the velocity, the formula being

$$p = \frac{wv^2}{2g},$$

where p is pressure in pounds per square foot, v is velocity of advance in feet per second, w is weight of a cubic foot of water and g is the acceleration due to gravity. If we assume that at a blade edge there is always a small portion which is virtually a plane surface, it follows that the motion of the blade through the water will cause at its edge the pressure given by the above formula.

Table XV shows for various blade edge velocities in feet per minute, g being taken as 32.16, the corresponding pressures in salt water weighing 64 pounds to the cubic foot. The pressures are expressed in pounds per square inch.

When we consider in Table XV the very rapid growth of blade edge pressures with velocity and the very high pressures reached when the velocity is 10,000 feet per minute and over, it is obvious that for high-speed propellers the area of blade edge over which such pressures are set up must be reduced to a minimum. In former days propeller blades were often made of elliptical section, and even now, for fairly high-speed propellers ogival blades are frequently finished with a quarter round. Such blades will certainly break down by cavitation at high-speeds and quick running propellers should by all means have sharp leading edges. It is difficult to make an edge which is mathematically a sharp edge, but the more nearly this is approached the better.

28. Wake Factor, Thrust Deduction, and Propeller Suction

Hitherto the ship and the propeller have been considered apart. It is necessary now to take up their very important reactions upon one another when the ship is being driven by its propeller or propellers.

1. **Components of Wake.** — Owing to its frictional drag upon the surrounding water there is found aft in the vicinity of the ship a following current or wake, called the frictional wake, which is in most cases greatest at the surface and in the central longitudinal plane of the ship and decreases downward and outward on each side. Superposed upon the frictional wake there is a stream line wake, caused by the forward velocity of the water closing in around the stern. This also will be greatest at the surface and center and decrease downward and outward, though its law of decrease will be different from that of the frictonal wake.

Superposed upon the two wakes above we have the wave wake. If there is a wave crest under the stern, the water is moving forward with velocity which decreases downward from the surface and, probably in practical cases, decreases slightly outward from the center.

Under a wave hollow the velocity is sternward,— the wave wake velocity in this case may be said to be negative,— the wake being regarded as positive when its velocity is forward.

There is a final factor, often ignored, which will be considered in more detail later in connection with shaft obliquity. The water aft is not flowing exactly parallel to the shaft. It rises up behind the stern and closes in horizontally, thus causing the slip of a propeller blade to be greater than the average over one portion of its revolution and less than the average over another. This condition of affairs does not materially affect the wake action, except in certain cases that will be considered later. For the present we will consider the wake proper — made up of the three components enumerated above.

2. Effects of Wake. — The propeller of an actual ship does not work in undisturbed water, but in water which has a very confused motion. The wake velocity varies over the propeller disc at a given speed, and at a given point of the disc varies with the speed. It is necessary to assume a uniform velocity of wake over the screw disc. This velocity of wake may conveniently be expressed as a fraction of the velocity of the ship, the ratio being called the "wake fraction" and denoted by w. The wake was first explored by R. E. Froude, who published some methods and results as long ago as 1883 in a paper before the Institution of Naval Architects. Froude used model propellers behind ships' models. Suppose the speed of the ship model is V. If the model screw is tested at given revolutions separate from the model at a speed of advance V into still water, we get a certain thrust and torque.

Suppose, now, keeping the revolutions constant, the model screw is tested behind the ship model. The thrust and torque are changed and are the same as would be found at the constant revolutions at a speed of advance V_1, say, into still water. V_1 is nearly always less than V. So the wake behind the model at the speed V is equivalent, so far as the screw is concerned, to a uniform following current of velocity $V - V_1$ or wV. The thrust and torque of the screw are then those appropriate to a speed of advance of V_1. The power absorbed is the same as if the screw were working in undisturbed water with speed of advance V_1. But if T denotes the thrust, the useful work as far as the ship is concerned is not TV_1 but TV. Hence the efficiency or ratio between the useful work and power absorbed is, if V is greater than V_1, greater than in undisturbed water, the ratio being $\dfrac{V}{V_1}$. The fact is that the following wake assists in pushing the ship ahead, using the propeller as the intermediary.

3. Thrust Deduction and Hull Efficiency. — While the ship acts upon the screw through its wake, the screw acts upon the ship through its suction.

Through its suction, the resistance of the ship is virtually increased beyond what it is without the screw. This is a cause of increase of power absorbed in propulsion. If R is the resistance of the ship at speed V, and T the screw thrust required to drive the ship at speed V, we have T greater than R. The quantity $T - R$ is called the thrust deduction, being the difference between the actual thrust and the net thrust or tow-rope resistance. It is usually denoted by tT, so that $R = T(1 - t)$ and $1 - t$ is called the thrust deduction factor, t being called the thrust deduction coefficient.

Suppose, now, we have a propeller absorbing a certain power, P, at certain revolutions per minute and driving a ship at speed V. In undisturbed water the propeller when absorbing the same power at the same revolutions would have a speed of advance V_1, and its efficiency would be a definite quantity, e say. Its thrust is T. Denote the effective horse-power necessary to propel the ship by E and its resistance by R. Then E is not equal to eP, as it would be if there were no wake or thrust deduction, but to $eP \times \dfrac{R}{T} \times \dfrac{V}{V_1}$.

The expression $\dfrac{R}{T} \times \dfrac{V}{V_1}$ is called the hull efficiency, and its two factors $\dfrac{R}{T}$ and $\dfrac{V}{V_1}$ are called respectively the thrust deduction factor and the wake factor. Since $R = T(1 - t)$ and $V_1 = V(1 - w)$ we have the hull efficiency $= \dfrac{R}{T} \times \dfrac{V}{V_1} = \dfrac{1 - t}{1 - w}$.

Froude expressed the wake as a fraction of V_1, the speed of advance, not V, the speed of the ship. Calling this w_p, Froude denoted the wake factor $\dfrac{V}{V_1}$ by $1 + w_p$ where w_p is the "wake percentage."

There are some advantages in using the "wake fraction" as already defined, but care must be exercised not to confuse it with Froude's "wake percentage." The relation connecting them is

$$w = \frac{w_p}{1 + w_p}.$$

In most cases the hull efficiency does not depart greatly from unity, the thrust deduction factor $1 - t$ being less than unity, and the wake factor $\frac{1}{1 - w}$ greater than unity.

This is readily understood when we reflect that the more favorably a screw is situated to catch the wake the more direct its suction as a rule upon the after part of the ship. Single screws, for example, may be expected to show larger thrust deductions and wake factors than twin screws. Also the stream line wake is increased by full lines aft, but the fuller the after part the stronger the propeller suction upon it and the larger the thrust deduction factor.

4. Variations of Wake Fraction and Thrust Deduction. — The wake fraction and thrust deduction are affected by many considerations, and in the present state of our knowledge the actual values in a given case can seldom be estimated accurately without special model experiments.

The most comprehensive information in this connection available at present is contained in a paper read at the 1910 Spring Meeting of the Institution of Naval Architects by W. J. Luke, Esq. This paper contains data as to the wakes and thrust deductions of models of various vessels that had been previously published, mainly by Mr. R. E. Froude, and gives a great deal of valuable new information obtained at the John Brown and Company's experimental tank at Clydebank, Scotland. These experiments were made with a single model 204 inches long, 30 inches broad, of 9 inches mean draught, displacement 1296 pounds in fresh water and having .65 block coefficient. All variations of propellers, etc., were tried with the bare hull and many with propeller bosses or brackets inclined $22\frac{1}{2}$ degrees from the horizontal. In addition some special experiments were made with bosses at other angles, ranging from horizontal to vertical.

In what may be termed the standard conditions, two three-bladed model propellers 6 inches in diameter, of 7.2-inch pitch with straight elliptical blades were used with centers $1\frac{1}{2}$ inches forward of the after perpendicular and 5 inches from the center line.

Experiments were made varying separately speed of vessel, pitch ratio and diameter of propellers, fore and aft and transverse position of propellers, number and area of blades, etc.

Briefly summarizing the main results of the twin screw experiments, which were always made with both outward and inward turning screws, Luke found that variation of number and area of blades had no appreciable effect upon wake factor and thrust deduction.

Change of pitch ratio produced changes of secondary importance for the bare hull, both wake and thrust deduction increasing slightly. With the $22\frac{1}{2}$ degrees bossing the changes were slight and much as before with outward turning screws, but with inward turning screws the wake fell off with increase of pitch.

Changes of diameter caused material changes in wake and thrust deduction, but Luke concluded that they were due as much to changes in clearance between hull and propeller as to the changes in diameter *per se*.

Change of speed of vessel resulted in practically no change in thrust deduction, but whether with bare hull or bossing the wake fell off steadily with increase of speed, the wake fraction decreasing with the bare hull and propellers in standard location from about .19 for speed-length ratio of .6 to .1452 for speed-length ratio of 1.0. In the paper the wake is characterized by the wake percentage values following Froude. These have been converted to wake fractions as already defined. For a speed-length ratio of .8, about what such a vessel would usually be driven at in service, the wake fraction was .167 for inturning screws and .173 for outturning screws, the thrust deduction t being about .155 in each case.

With the bossing the thrust deduction was still practically the same with out- and inturning screws and varied little from .16. The wake fraction fell off with the speed as with the bare hull, but the wake was materially greater for outturning than for inturning screws. For the .8 speed length ratio it was .191 instead of

.173 for outturning screws and .146 instead of .167 for inturning screws.

Luke's experiments show clearly that for the model tried the most important factor affecting wake and thrust deduction is the location of the propeller with reference to the hull. Thus with the bare hull and the 6-inch propeller the results were as follows:

Center of propeller from center of model	4″	5″	6″
Wake fraction — outturning screws	.238	.171	.119
Wake fraction — inturning screws	.198	.167	.131
Thrust deduction, t, both cases	.166	.150	.116

It is seen that in this case a transverse change of $\frac{1}{3}$ the diameter caused wake and thrust deduction to vary a great deal, both being larger the closer the propeller was to the hull. When the distance of the propeller from the hull was varied by shifting it fore and aft the effect was not so great, but still material.

A few experiments were made with a single screw behind the model, diameter being varied. The wake was found markedly greater for this propeller location, the thrust deduction being also increased, but not nearly so much as the wake. Curiously enough the smaller the propeller the larger the wake. Thus, the wake fraction varied from about .275 for a 5-inch screw to .226 for an 8-inch screw. The corresponding thrust deduction values were .155 and .185, the smaller screw thus profiting not only by the larger wake but by the smaller thrust deduction.

Luke's paper makes it clear that location with respect to the hull is a very important factor in connection with wake and thrust deduction. Experiments such as described in his paper made with models of varying fineness are much needed.

5. **Approximate Wake Fractions and Thrust Deductions.** — Since the speed of advance of a propeller, a vital factor in design, depends upon the wake fraction, it is important to be able to approximate to it in a given case. In Luke's paper, as already stated, are given a number of wake factors for single and twin screw ships and of thrust deduction coefficients, t, for twin screw vessels. For twin screw vessels, Froude laid down the dictum many years ago that,

broadly speaking, wake factor $\dfrac{1}{1-w}$ and thrust deduction factor $1 - t$ were reciprocals or $w = t$. The data given by Luke confirms this, and shows also that we may, so far as present knowledge goes, reasonably assume wake fraction to vary linearly with block coefficient.

Then from the data published by Luke we may say with reasonable approximation $w = -.2 + .55\,b = t$, where w is wake fraction, t is thrust deduction coefficient and b is block coefficient. This formula ignores the matter of screw location, but may be taken as applying to screws about abreast the after perpendicular and with centers about 1.2 the radius from the center line.

For lesser clearance w will be greater and t will also increase somewhat, but the formula is and can be, from the available data, only a rough approximation.

For center screws in the usual position the approximate formula indicated is $w = -.05 + .5\,b$.

Data is not available for a formula for t for center screws, but Luke's experiments would appear to indicate that for them t would be increased but little over its value for twin screws. It follows that if the hull efficiency is unity for twin screws it is somewhat over unity for single screws, particularly for full vessels.

The formulæ above apply to the bare hull or to vessels fitted with struts or bossing which does not interfere with the natural water flow.

It should be remembered that they are deduced from model experiments and will nearly always exaggerate the wake of the full-sized ship. It is desirable, however, if we cannot determine the wake accurately, to overestimate it rather than underestimate it. If it is overestimated, the engines on trial will turn somewhat faster than estimated, which is generally allowable. If it is underestimated, it may be impossible to run the engines up to the designed speed without decreasing propeller pitch or reducing propeller diameter.

6. Approximate Determination of Wake Fraction. — Since the wake is explored by trial of model screws working behind models of ships the question naturally arises whether we cannot gain some light upon the subject from trials of full-sized ships. Analysis soon makes it evident that the apparent slip of propellers on trial is often very much below what must have been the real slip. We know that in any case the power absorbed by a given propeller advancing through undisturbed water depends only upon the revolutions and the speed of advance. For an actual propeller advancing through the water disturbed by the ship we can reasonably reduce the actual disturbance to an equivalent uniform motion. Throughout the range where the Law of Comparison holds we can determine for any propeller for which we have model experiments the relations between power absorbed, revolutions and speed of advance. Hence, if we know any two of these quantities, we can determine the third. Now from the results of trial of a vessel we know corresponding values of indicated or shaft horse-power, revolutions and speed of vessel. The shaft horse-power is practically the power, P, absorbed by the propeller, and from the indicated horse-power P can be estimated with reasonable accuracy. Hence, although we do not measure V_A directly, we can estimate it from the power and revolutions if we have reliable model experiments with the propeller and the Law of Comparison holds, and knowing V_A and V we can determine the wake fraction. The reduction of the results of model experiment to a form convenient for this application is simple. We have seen that we may write $P = A \dfrac{d^2 V_A^3}{1000}$ where P is power absorbed by the screw, d is diameter in feet and V_A is speed of advance in knots. A is a coefficient independent of size and speed and depending only upon the slip and the proportions and shape of the propeller.

So let us write $P = S \left(\dfrac{pR}{1000} \right)^3 \dfrac{d^3}{p}$ where p is pitch in feet and R denotes revolutions per minute. Then $S = \left(\dfrac{1000}{pR} \right)^3 \dfrac{p}{d^3} P$ and is like A, a coefficient independent of size and speed and depending only on the slip and the proportions, etc., of the propeller.

From experimental results with models we can readily determine a curve of S plotted on the slip. Thus, for a 16-inch model with a speed of advance of 5 knots we have $S = .3129\, Q\, (1 - s)^2$ where Q

is torque in pound-feet. Or we may determine S from a curve of A.

For $V_A = \dfrac{pR(1-s)}{101.33}$. So $P = Ad^2 \dfrac{(pR)^3(1-s)^3}{1000\,(101.33)^3} = S\left(\dfrac{pR}{1000}\right)^3 \dfrac{d^3}{p}$.

Whence $S = .9610 \dfrac{p}{d} A (1-s)^3$.

Fig. 229 shows curves of S plotted on s for the four propellers of Fig. 179. Now suppose we have a full-sized propeller similar to the model of .0448 blade thickness fraction and 18 feet in diameter, making 120 revolutions per minute and absorbing 12000 horsepower. Its pitch will be 21.6 feet. Then from the data of the full-sized screw $S = \left(\dfrac{1000}{pR}\right)^3 \dfrac{p}{d^3} P = 2.552$. From Fig. 229, for the propeller in question, when $S = 2.552$, $s = .2340$. So the true slip of this propeller would be .2340, and its true speed of advance, $V_A = \dfrac{pR(1-s)}{101\frac{1}{3}} = 19.593$. Suppose the speed of the ship V is 21 so that the apparent slip, s', is .1790.

Then $V = \dfrac{pR(1-s')}{101\frac{1}{3}} = 21$ knots.

The wake $= V - V_A = \dfrac{pR}{101\frac{1}{3}}(s - s') = 1.407$ knots.

Wake fraction $= \dfrac{V - V_A}{V} = \dfrac{\dfrac{pR}{101\frac{1}{3}}(s-s')}{\dfrac{pR}{101\frac{1}{3}}(1-s')} = \dfrac{s - s'}{1 - s'} = .0670$.

It is very easy to derive curves of S from the Standard Series results of Figs. 185 to 208.

Figures 230 to 233 show contours of slip plotted on S and pitch ratio for four blade widths and the blade thickness fractions indicated. For propellers closely resembling the Standard Series these figures may be used in connection with accurate trial data to obtain a reasonable approximation to the wake so long as there is no cavitation. The propeller power, P, however, must for reciprocating engines be estimated from the I.H.P. Methods for this will be considered under analysis of trials.

These figures may be used, however, to obtain rough approximations to the wake for propellers very different from the Standard Series.

For three-bladed propellers with oval blades and extra wide tips the correct values of S will be somewhat less than in the figures, but the difference for practical propellers will not be great. In order to use Figs. 230 to 233 for four-bladed propellers we need only divide the actual propeller power, P, by the proper power ratio for four blades, obtained from Fig. 181. We thus obtain approximately the power absorbed by a three-bladed propeller having blades identical with the four-bladed propeller and working with the same revolutions and speed of advance.

From this we determine S and use Figs. 230 to 233 as before. It will be found in practice that the methods above for estimating the wake from full-sized trials will generally give values that seem too low. We know that the wake values for a full-sized ship should be less than for its model, but another factor present at times and tending to lower the wake deduced from the S value is a slight failure of the Law of Comparison connecting model and full-sized propeller. We know that the Law of Comparison fails when a propeller breaks down by cavitation, but it is probable, particularly with blunt-edged blades, that there is more often than might be supposed a certain amount of eddying in the operation of the full-sized propeller not found in the operation of the model. This might not seriously reduce efficiency and would manifest itself mainly by a slip of the full-sized propeller somewhat larger than would be inferred from the model results. The wake deduced from the S values would be correspondingly reduced.

The S value method should not be used when the wake can be investigated by model experiments. Lacking model experiments, we can roughly approximate to the wake by the formulæ already given.

There is great need for a systematic and thorough experimental investigation of the question of wake, following the lines of Luke's experiment, which will enable it to be closely estimated in any practical case likely to arise. But there is a mass of accummulated trial data extant for vessels whose models never have been, and probably never will be, tested, and it is worth while for those possessing it to investigate the wake fraction even by a method which is only roughly approximate. For practical purposes the wake fraction of a vessel seldom requires to be determined with minute accuracy. It is principally of use for settling the diameter and pitch of the screw, and neither these nor the efficiency will often be much affected by a moderate error in the wake fraction.

If by use of Figs. 230 to 233 we find a certain wake for a vessel of a given type, we can use this for a vessel of the same type with similar propeller location, and for the purpose of determining diameter and pitch of screw it will make little difference whether the nominal wake from Figs. 230 to 233 is the real wake or departs materially from it. Whatever the departure, it will be practically the same in the two cases.

7. **Effect of Shaft Brackets upon Wake.** — Reference has already been made to the apparent effect upon the wake of the direction of flow of the water aft.

This has a marked effect when large shaft brackets are fitted which modify the natural flow of the water.

Thus, if a shaft bracket is fitted with a wide horizontal web, it interferes seriously with upward flow aft and the water closes in with a much stronger horizontal motion or current inwards than otherwise. The conditions over the lower half of the propeller disc are somewhat, but not very seriously, modified from bare hull conditions, much greater modifications occurring over the upper half of the disc. Considering the upper blades, the effect of the inward flow of the water is materially to increase the slip angle for outward turning propellers where the upper blades are moving against the current, while for inward turning screws with the upper blades moving in the same direction as the current the slip angle would be decreased. Hence, we may expect a large horizontal shaft bracket materially to increase the apparent wake for outward turning screws and to decrease it for inward turning screws.

A case in point is that of the *Niagara II*, a steam yacht 247′ 6″ × 36′ × 16′4½″ draught and 2000 tons displacement.

This vessel had a Lundborg stern, involving wide horizontal shaft brackets, and her deadwood aft was not cut up.

She had two six-hour trials under similar conditions, except that the screws were interchanged, being inward turning on the first trial and outward turning on the second. While the horse-power was not accurately determined, it was closely estimated at 2100 with inward turning screws and 1950 with outward turning screws.

Nevertheless, with inward turning screws the average speed was 12.8 knots with an apparent slip of 26.4 per cent, while with outward turning screws the average speed was 14.12 knots with an apparent slip of but 13.3 per cent.

This marked difference in apparent slip can be due only to the fact that the horizontal shaft webs force a strong inward motion of the water above them along horizontal lines, and while this motion is not a wake, being transverse or perpendicular to the line of advance of the ship, its effect upon the upper blades of the propeller is equivalent to a positive wake for outturning screws and a negative wake for inturning screws.

It would seem that the lower blades are not much affected, such action as there may be upon them being much less than that upon the upper blades.

Luke's paper already referred to, gives most interesting and instructive results of a model investigation of shaft bracket angles and direction of screw rotation. The model was the same as already described, 204 inches long, 30 inches broad, 9 inches draught, 1296 pounds displacement in fresh water, having a block coefficient of .65. The model screws were three-bladed, 6 inches in diameter, 7.2 inches pitch, having straight elliptical blades. Their centers were 5 inches out from the center line of the model and $1\frac{1}{2}$ inches forward of the A.P. Brackets were fitted at angles ranging from horizontal to vertical and the model tested with inturning and outturning screws, the screws and their positions remaining unchanged as the shaft bracket angles were varied. The results are summarized on the following page.

These results show relatively enormous variations of wake with variation of bracket angle and direction of turning and make it clear that under some conditions the virtual wake due to obliquity of water motion may overshadow the real wake or forward motion. It is obvious that for a given real wake outturning and inturning screws should give practically the same derived wake. We see, however, that with horizontal brackets the wake fraction is about $2\frac{1}{2}$ times as great with outturning screws as with inturning screws while with vertical brackets the wake fraction with inturning screws is nearly four times as great as with outturning screws.

	Hori- zontal.				Verti- cal.
Angle of bracket with horizontal................	0°	$22\frac{1}{2}$°	45°	$67\frac{1}{2}$°	90°
Wake fraction — outward turning screws......	.243	.190	.145	.099	.065
Wake fraction — inward turning screws.......	.099	.145	.184	.221	.254
Thrust deduction — outward turning screws...	.164	.169	.173	.184	.215
Thrust deduction — inward turning screws ...	.155	.160	.164	.171	.189
Hull efficiency — outward turning screws......	1.105	1.026	.967	.906	.840
Hull efficiency — inward turning screws	.938	.982	1.025	1.064	1.087
Model resistance in terms of bare hull resistance	1.094	1.040	1.028	1.052	1.120

These differences can be due only to the fact that transverse motion of the water affects inturning and outturning screws very differently. Horizontal motion inward is equivalent to a positive wake for outturning screws and a negative wake for inturning screws. Vertical motion upward is equivalent to a negative wake for outturning screws and a positive for inturning screws.

In the light of Luke's experiments the remarkable trial results of the *Niagara II* are readily explicable.

While horizontal shaft brackets in his experiments resulted with outturning screws in a high hull efficiency this was accompanied by an increased hull resistance, so there was no appreciable net gain.

It would seem that in practice from the point of view of resistance and propulsion, shaft brackets should be arranged to parallel if possible the lines of flow. It is generally, however, more convenient to arrange them more nearly horizontal and when this is done the screws should obviously be outturning.

8. Propeller Suction. — The thrust deduction is due to the suction of the propeller upon the ship's hull. It is well to consider in connection with it the question of propeller suction generally and its effect upon the water. An experimental investigation in this connection has been made at the United States Model Basin, and is described in a paper entitled " Model Basin Gleanings " read before the Society of Naval Architects and Marine Engineers in 1906. The suction of 16-inch model propellers was measured over the surface of a vertical plane, parallel to the propeller axis, which could be set at various distances from the propellers. Figure 234, showing the variation of pressure along various horizontal lines of the plane when set $\frac{3}{8}$ inch from the tips of a propeller of 16-inch diameter and 16-inch pitch which was working at a nominal slip of 30 per cent, is typical of all the results.

Now a necessary result of this suction is that it draws the water inward toward the propeller axis and aft toward the disc. An important fact, which seems to have been generally ignored, should be pointed out. When a propeller works with sternward slip velocity of the water, the supply of water necessary to allow slip velocity comes ultimately from the free surface. For referring to Fig. 235, which indicates a submerged propeller, consider an imaginary plane XY perpendicular to the shaft axis and just forward of the screw disc, as indicated by the dotted line. But for the screw action all the water in that plane would be at rest. Owing to the screw action the water is flowing aft through the screw disc and forward is flowing from all directions toward the disc. Now the water flowing through the plane does not leave a vacuum behind it; and particles of water flowing toward the disc from points forward of the plane cannot leave vacua behind them. Their places must be taken by other particles of water. Where can these particles come from? The water being practically incompressible, there are only two possible sources of supply. It is possible to conceive that the water flowing aft through the plane just forward of the screw disc spreads out astern, and finally to an equal amount flows forward again through the plane. In other words, the suction draws a certain amount of water through the plane and the thrust behind the propeller forces an equal amount across the plane in the opposite direction at points some distance from the disc. This action goes on when a screw is operated with no speed of advance as in dock trials. Careful study of the action of advancing screws, however, indicates clearly that in this case the water to take the place of that sucked to the propeller disc simply flows downward from the surface, producing a depression of the surface, which advances with the speed of the propeller. Figures 235, 236 and 237 show results of an experimental investigation of this question made at the United States Model Basin. Two 16-inch propellers of identical blade profiles, as indicated, one with 12.8-inch nominal pitch, the other with 19.2-inch nominal pitch, were operated as indicated with their tips 8 inches below the surface and the resulting surface depressions for 5-knots speed of advance and various slips observed. It is seen that contour lines in the depression over the propeller are approximately circular. The point of maximum depression is in each case a little astern of the propeller, and as to be expected, the greater the slip the greater the depression; also the finer the pitch the greater the depression. This too is to be expected. The propeller of fine pitch exerts much the greater thrust for a given slip and speed of advance. Hence the actual sternward velocity communicated to the water is greater for the propeller of fine pitch than for the propeller of coarse pitch, and the surface depression greater accordingly.

As a result of the fact that sternward velocity of water entering the screw disc is obtained ultimately by sucking water from the surface, it follows that if a screw is so arranged that it cannot draw water from the surface, the sternward velocity of the water entering the screw disc is reduced. The suction in such cases, not being absorbed by giving velocity to the water, is likely to be exerted upon the ship and cause abnormal thrust deduction. Once the water has reached the screw disc it is difficult to conceive, as pointed out in discussing Rankine's theory, how it can be given much additional sternward velocity. We must conclude that while in the disc the change of velocity is nearly all rotary, as in Greenhill's theory. It is true that this involves changes in pressure, and Greenhill, on account of the increase of pressure involved in his theory, considered it necessary to confine the screw disc and race by a cylinder. Greenhill has pointed out, however, that it is conceivable to have a defect of pressure behind the screw at the center, the pressure increasing as the circumference is approached until at the outside of the screw race it is normal. It should be pointed out that, since there is quite a defect of pressure in all the water passing into the screw disc, its pressure while in the disc can be materially increased by the action conceived by Greenhill without exceeding the normal pressure of the surrounding water.

To sum up, it appears that a reasonable theory of what happens to a particle of water which is acted on by a propeller is about as follows: When some distance forward of the screw, it is sucked aft and in toward the shaft axis, its pressure being reduced at the same time. Hence, it enters the screw disc with a certain sternward velocity and reduction of pressure. As it passes through the disc its sternward velocity is changed but little. It has impressed upon it a rotary velocity and an increase of pressure, so that its pressure on passing out of the screw disc is probably very close to normal pressure again for particles near the circumference of the screw race and still below normal for particles in the interior of the race.

9. **Effect of Immersion upon Suction and Efficiency.** — The sternward velocity into the screw disc is affected by the situation of the screw. Probably immersion alone does not affect it much. The more deeply immersed screw is, it is true, farther from the surface from which its water supply must come, but it is in a position to draw upon a larger surface area. Still from this point of view there is nothing favorable to efficiency in deep immersion, the reasons rendering it desirable in most cases and necessary in some having to do not with efficiency but with prevention of racing in a seaway.

If vessels worked always in smooth water, there is little doubt that screws could be located with their tips quite close to the surface, provided they did not suck air in operation, without loss of efficiency. In fact, in a paper by W. J. Harding, read March 13, 1905, before the Institute of Marine Engineers, on "The Development of the Torpedo Boat Destroyer," we find the statement when discussing the question of propellers of destroyers:

" The least immersion of the propellers gave the best results, both in speed and coal bill." This conclusion was deduced from consideration of a number of trial results of destroyers in smooth water.

A screw propeller placed under a wide flat stern, or with the flow of water to it obstructed in any way by the hull to which it is attached, must evidently work more after the Greenhill theory than a screw with a free flow of water to it.

Apart from the increased thrust deduction this must involve a reduction of propeller efficiency. It is, of course, necessary at times to fit screws in tunnels, or so that they are hampered by the hull, but when this must be done allowance should be made for the loss of efficiency involved.

29. Obliquity of Shafts and of Water Flow

1. **Shaft Deviations, Actual and Virtual.** — Propeller designs and calculations are usually based explicitly or implicitly upon the assumption that the propeller advances in the line of the shaft axis. As a matter of fact, it is unusual to find a shaft which is exactly horizontal when the propeller is working. Shafts of center screws are in a fore and aft line, but side screw shafts generally depart in plan from the fore and aft line.

The divergence of propeller shafts from a horizontal fore and aft line is seldom so great that the resolved horizontal fore and aft thrust differs materially from the axial thrust. But there is a very serious departure from ideal conditions as regards slip of blade during revolution. The slip angle is a small angle, as a rule, and if the shaft axis is changed from the line of advance of the screw, the slip angle at one part of the revolution is increased by the amount of angular change and at another part is decreased by an equal amount. The slip angle is a function of the slip ratio and the pitch ratio or diameter ratio. Fig. 170 shows slip angles for the range of pitch ratio and slip ratio found in practice.

The small size of these slip angles renders it evident that shaft deviations occurring in practice must cause the slip of a blade to vary materially during a revolution.

2. **Wake and Obliquity of Water.** — The variation of wake is another perturbing factor. The slip of the blade will be greatest where the wake is strongest. Evidently a virtual deviation of shaft axis can be imagined which would give practically the same effect as the variation of wake. Finally, the water itself has a motion across the shaft axis.

3. **Variation of Slip.** — The net result is that, in practice, instead of the thrust, torque and efficiency of a blade remaining constant during a revolution, they vary throughout the whole revolution. To fix our ideas, suppose we consider a starboard side propeller turning outward. In considering shaft inclination we will always take it as we proceed forward from the propeller.

If the shaft inclines upward from the propeller, the slip angle will be decreased by the amount of shaft inclination for a blade in a horizontal position inboard and increased by the same amount for the blade in a horizontal position outboard. For the blade at the top and bottom of its path there will be no appreciable change. Similarly, for a shaft inclined inboard, as we go forward there will be no effect for the horizontal position of the blades, a maximum increase of slip for the top position of the blade and a maximum decrease for the lower position of the blade.

If the wake is strongest next the hull on a horizontal line, the result is equivalent to a downward inclination of the shaft, hence we may say that such a wake causes a virtual downward inclination. Similarly, a wake strongest nearest the surface gives a virtual inclination inward. Water rising up gives a virtual upward inclination, and water closing in gives a virtual inward inclination.

The table below gives the positions for maximum and minimum slip of blades due to shaft inclination. Of course, when the shaft has both horizontal and vertical inclination, the positions of maximum and minimum slip are neither horizontal nor vertical. In all cases, the plane of zero effect is that including the shaft axis and the line of advance of the center of the propeller. The plane of maximum effect is that through the shaft axis perpendicular to the preceding.

BLADE POSITIONS OF MAXIMUM AND MINIMUM SLIP DUE TO SHAFT INCLINATIONS RECKONED FROM PROPELLER FORWARD.

Shaft Inclination.	Right-handed Screws.						Left-handed Screws.					
	Port.		Center.		Starboard.		Port.		Center.		Starboard.	
	Maximum Slip.	Minimum Slip.	Maximum Slip.	Minimum Slip.	Maximum Slip.	Minimum Slip.	Maximum Slip.	Minimum Slip.	Maximum Slip.	Minimum Slip.	Maximum Slip.	Minimum Slip.
Up.......	in	out	S	P	out	in	out	in	P	S	in	out
Down....	out	in	P	S	in	out	in	out	S	P	out	in
Inboard..	down	up	....	....	up	down	up	down	....	....	down	up
Outboard	up	down	....	....	down	up	down	up	....	....	up	down

In the above, "in" means that the blade is in the horizontal position next the ship. "Out" means that the blade is in the horizontal position away from the ship. P means horizontal position to port for center screw and S the horizontal position to starboard. "Up" means blade vertical upward, "down" means blade vertical downward.

The following table gives virtual inclinations of shaft corresponding to wake and transverse motions of the water:

TABLE OF VIRTUAL SHAFT INCLINATIONS FOR MOTION OF WATER INDICATED.

Motion of Water.	Right-handed Screws.			Left-handed Screws.		
	Port.	Center	Starboard	Port.	Center	Starboard
Wake increasing inward.........	up	......	down	down	......	up
Wake increasing upward.........	out	......	in	in	......	out
Water rising vertically...........	up	up	up	up	up	up
Water closing in horizontally.....	in	......	in	in	......	in

4. Virtual Inclinations in Practice. — In practice, in most cases of twin screws, the wake increases inward and upward and the water rises vertically and closes in horizontally, the latter motion being strongest over the upper half of the disc. Then for inturning screws (port right-handed and starboard left-handed) we have a positive virtual upward inclination, since both water motions give a virtual upward inclination. As regards horizontal angle the wake gives a virtual outward inclination and the horizontal water motion a virtual inward inclination. So the net virtual inclination may be either in or out, being the difference of the two components.

For outturning screws we have a positive virtual inward inclination, and vertically the inclination is the difference of two virtual inclinations.

If we wish to secure uniform turning force on each blade, we must neutralize the virtual shaft inclination due to water motion by actual inclination in the opposite direction.

While we have not data for exact quantitative results, it is evident from what has been said that with inturning screws the shafts should incline downward from the screws and for outturning screws the shafts should incline outwards.

Outturning screws, with shafts inclining outward, are desirable for maneuvering purposes.

5. Effect upon Efficiency. — This question of desirable shaft angles is of importance in practice, and it is to be hoped that some day it will be given accurate experimental investigation. At present we can deal with it in quantitative fashion only. As regards efficiency, a moderate variation of slip during the revolution of a blade will not seriously reduce efficiency so long as the average slip is that corresponding to good efficiency and the variation of slip is not extreme. But it is difficult to see how a shaft inclination as great as ten degrees, which has been often fitted on motor boats, can fail to be accompanied by a loss of efficiency. With such an angle of inclination it is evident, from Fig. 170, that each blade will work with negative nominal slip at one portion of its revolution and with excessive nominal slip at another portion — even if the average slip is that corresponding to good efficiency.

If the thrust of a propeller were due solely to the action of the face such a variation of slip would be wholly inadmissible. Irregular turning forces and thrust would cause serious vibration and there would be great loss of efficiency. But the back of the blade through its suction is always an important and often a dominant factor in the production of thrust. The slip angle for the following portion of the blade is greater than the slip angle for the face by the value of the edge angle at the following edge. This edge angle is seldom less than twelve or fifteen degrees and is often twenty-five or thirty. Hence a shaft inclination of two or three degrees will affect comparatively slightly the action of the blade back, and even the large inclination of ten degrees will seldom cause the suction of the back to be reversed into negative thrust at any portion of the revolution. Such a large deflection, however, is liable to produce very irregular action.

6. Vibration. — An important consideration in this connection is that of vibration. With turbine propelled vessels, practically all vibration — which is quite strong in some turbine steamers —

is due to pounding of the water against the hull as the blades pass or to unbalanced propeller action. There can be no doubt that the latter cause of vibration, which is practically the only cause if the propeller tips are not too close to the hull, is affected by the shaft angles, and it is particularly advisable with turbine steamers to choose shaft angles which will tend to uniformity of propeller action. Suppose, for instance, we have a propeller shaft carried by a nearly horizontal web. We have seen that there will be a very strong wake above the web and vertical motion of the water will be interfered with. In such a case, for inturning screws, the shaft should incline down and out, and for outturning screws, up and out.

7. **Obliquity of Flow.** — While the wake through its variation of strength over the propeller disc produces a virtual shaft deviation, it is evident from consideration of Figs. 50 to 59, showing lines of flow over models, that the water closing in and rising up aft follows lines which will in many cases make material angles with the shafts. The effect of the obliquity of the water flow will vary a good deal with the position of the propeller.

For vessels of usual type it would seem that the farther aft the propeller the less the obliquity of the water flow. But experiments with the model of a four-screw battleship indicated that at the forward screws the water was rising at an average angle of about $10°$ and closing in at an average angle of about $5°$. For the after screws these angles were $11°$ and $4°$ respectively. The after screws, however, were not very far aft. These angles seem large when we compare them with the slip angles to be expected in practice. The obliquity of horizontal water flow will usually be greater over the upper portion of the propeller disc than over the lower, so that the virtual wake to which the obliquity of motion is equivalent will be stronger over the upper portions of the screw disc. Now this virtual wake will, for outturning screws, be positive over the upper part of the disc and negative over the lower. For inturning screws the virtual wake will be negative over the upper portion of the screw disc and positive over the lower.

The strength of the virtual wake being in the upper part of the disc, where it is positive for outturning screws and negative for inturning, it would seem that side screws well forward of the stern post should be outturning in order to make the most of the virtual wake due to the obliquity of the water motion.

30. Strength of Propeller Blades

In view of the importance of blade thickness in many cases it is advisable to make a careful inquiry into the matter and endeavor to reduce to rule the stresses upon propeller blades. This can be accomplished only by certain assumptions, which will be pointed out and justified as they are made. In order to apply the well-known formula for beam stress to a propeller blade, it will be assumed that the section of a blade by a cylinder at a given radius is developed into a plane tangent to the cylinder. This section will then be treated as a beam section. This assumption probably errs on the safe side, since the actual strength as a beam of the curved blade would be greater than that of a developed cylindrical section of the same.

1. **Fore and Aft Forces and Moments.** — In considering the forces upon a blade it is convenient first to consider separately fore and aft forces, or thrust and transverse forces, producing torque. It is convenient to use the disc theory or Rankine's theory, by which the thrust upon a blade may be taken to vary radially directly as the distance from the shaft center. For a ring of water one inch thick at ten feet radius, say, would contain twice as much water as a ring of the same thickness at five feet radius. If each ring be given the same sternward velocity, involving the same thrust per pound of water acted upon, then the thrust from the ring of ten feet radius would be double that from the ring of five feet radius. Put into symbols, if dT denote elementary thrust from a ring of thickness dr at radius r, we have $dT = kr\,dr$ where k is a constant coefficient over the blade depending upon the total thrust. Then integrating we have for thrust,

$$T = \tfrac{1}{2} kr^2.$$

Applying the limits $\frac{d_1}{2}, \frac{d}{2}$, where d_1 is diameter of hub and d is diameter of propeller, we have, if T_0 is total thrust of one blade,

$$T_0 = \frac{k}{2}\left(\frac{d^2}{4} - \frac{d_1^2}{4}\right).$$

This enables us to determine k, since from the above

$$k = \frac{8\,T_0}{d^2 - d_1^2}.$$

Suppose, now, we wish to determine the thrust T_1 from the tip to a radius r_1,

We have
$$T_1 = \frac{k}{2}\left[\frac{d^2}{4} - r_1^2\right]$$
$$= \frac{4\,T_0}{d^2 - d_1^2}\left[\frac{d^2}{4} - \frac{d_1^2}{4} + \frac{d_1^2}{4} - r_1^2\right]$$
$$= T_0\left[1 - \frac{4\,r_1^2 - d_1^2}{d^2 - d_1^2}\right].$$

From the above, if $r_1 = \frac{d_1}{2}$, then $T_1 = T_0$, and if $r_1 = \frac{d}{2}$, then $T_1 = 0$ as it should. Now we need to know not only the thrust on the blade beyond any radius, but its moment at the radius. The moment at radius r_1 of the elementary thrust dT at radius r is $dT\,(r - r_1) = kr\,(r - r_1)\,dr$.

Call dM_1 the moment of this elementary thrust. Then

$$\frac{dM_1}{dr} = k\,(r^2 - r_1 r) = \frac{8\,T_0}{d^2 - d_1^2}(r^2 - r_1 r).$$

Then
$$M_1 = \frac{8\,T_0}{d^2 - d_1^2}\left[\frac{r^3}{3} - \frac{r_1 r^2}{2}\right]_{r_1}^{\frac{d}{2}}.$$

Upon reduction we have

$$M_1 = \frac{T_0}{3\,(d^2 - d_1^2)}\,(d^3 - 3\,r_1 d^2 + 4\,r_1^3).$$

At the root section $r_1 = \frac{d_1}{2}$. Substituting and reducing, we have at the root section,

$$M_1 = T_0\,\frac{2\,d^2 - d d_1 - d_1^2}{6\,(d + d_1)}.$$

Suppose, now, the thrust were concentrated at a point $k_1\,\frac{d}{2}$ out. Then we should have

$$M_1 = T_0\left(k_1\,\frac{d}{2} - \frac{d_1}{2}\right).$$

Whence, equating these two values of M_1, we have

$$k_1 \frac{d}{2} - \frac{d_1}{2} = \frac{2\,d^2 - dd_1 - d_1^2}{6\,(d + d_1)}.$$

Upon reduction this gives us

$$k_1 = \frac{2\,(d^2 + dd_1 + d_1^2)}{3\,d\,(d + d_1)}.$$

The value of k_1 in the above formula depends only upon the ratio between d_1, the diameter of hub, and d, the diameter of propeller. Numerical values are given below:

$d_1 =$	$\frac{d}{10}$	$\frac{2\,d}{10}$	$\frac{3\,d}{10}$	$\frac{4\,d}{10}$
$k_1 =$	.673	.689	.713	.743

These values of k_1 agree very well with values deduced by entirely different methods upon the blade theory or Froude's theory. Upon the blade theory k_1 is nearly constant at .7.

2. Transverse Forces and Moments. — Let us now take up the transverse moment, which denote by M_2. Let dQ denote the elementary transverse force in pounds at radius r. Let p denote pitch in feet, s the slip ratio and e the efficiency of the elementary portion of the blade at radius r. Then the gross work done by the element of the blade in one revolution is in foot-pounds $dQ \times 2\,\pi r$.

The useful work is

$$dQ \times 2\pi r \times e = dT \times p\,(1 - s) = krdrp\,(1 - s).$$

Whence
$$\frac{dQ}{dr} = \frac{kp\,(1 - s)}{2\,\pi e}.$$

Now over a blade the quantities on the right in the above equation are all constant except e. The variation in e over the part of the blade that does the most work is probably not great, so let us assume it constant and write $\frac{dQ}{dr} = g$, where g is a constant coefficient to be determined.

We have seen that $dQ \times 2\,\pi r =$ element of work done in one revolution in foot-pounds. Then $\int dQ \cdot 2\,\pi r =$ work per blade per revo-

lution $= \dfrac{33000\, P_1}{R}$, where P_1 is the power absorbed by one blade.
Then

$$\dfrac{33000\, P_1}{R} = \int dQ \cdot 2\pi r = \int g \cdot 2\pi r \cdot dr = \left[\pi g r^2\right]_{\frac{d_1}{2}}^{\frac{d}{2}}.$$

$$= \dfrac{\pi}{4} g (d^2 - d_1^2) \quad \text{or} \quad g = \dfrac{4 \times 33000}{\pi} \dfrac{P_1}{R} \dfrac{1}{d^2 - d_1^2}.$$

$$= 42{,}017 \dfrac{P_1}{R} \dfrac{1}{d^2 - d_1^2}.$$

Then for M_2, the transverse moment at any radius, r_1, due to the moments of the elementary transverse forces from the tip in to the radius, r_1, we have

$$M_2 = \int dQ\,(r - r_1) = g\int (r - r_1)\, dr = g\left[\dfrac{r^2}{2} - r r_1\right]_{r_1}^{\frac{d}{2}}$$

$$= \dfrac{g}{2}\left(\dfrac{d}{2} - r_1\right)^2.$$

Then, upon substituting its value for g and reducing, we obtain

$$M_2 = 5252 \dfrac{P_1}{R} \dfrac{(d - 2r_1)^2}{d^2 - d_1^2}.$$

Now as to the radial position of the transverse center of effort we have the total transverse force equal to

$$\int_{r_1}^{\frac{d}{2}} dQ = g \int_{r_1}^{\frac{d}{2}} dr = g\left(\dfrac{d}{2} - r_1\right).$$

The arm of this force beyond r_1 is obtained by dividing moment by total force and equals

$$\dfrac{\dfrac{g}{2}\left(\dfrac{d}{2} - r_1\right)^2}{g\left(\dfrac{d}{2} - r_1\right)} = \dfrac{1}{2}\left(\dfrac{d}{2} - r_1\right).$$

The center of transverse effort, then, is by this method halfway between the tip and the radius considered. So if $k_2 \dfrac{d}{2}$ denote the distance of the center of effort of the whole blade from the center of the propeller d_1, denoting the root diameter, we have

$$k_2 \frac{d}{2} = \frac{1}{2} d_1 + \frac{1}{2}\left(\frac{d}{2} - \frac{d_1}{2}\right),$$

or
$$k_2 = \frac{1}{2}\left(1 + \frac{d_1}{d}\right).$$

This gives us the values below of k_2 for the values of $\frac{d_1}{d}$ indicated,

$\frac{d_1}{d} =$	.1	.2	.3	.4
$k_2 =$	.55	.6	.65	.7

These compare fairly well with values of k_2 deduced by entirely different and more complicated methods upon the blade theory. These values of k_2 varied from .710 for a coarse pitch ratio of 2 to .600 for a pitch ratio of 1.

Let us now recapitulate the results to this point.

Let d denote the diameter of the propeller in feet,
       d_1 the diameter of the hub or diameter to root section,
       r_1 the radius to the point at which we wish to determine thickness,
      T_0 whole thrust of the single blade in pounds,
      P_1 horse-power absorbed by the single blade,
       R revolutions per minute,
      M_1 fore and aft bending moment at radius r_1 in lb.-ft.
      M_2 transverse bending moment at radius r_1 in lb.-ft.

Then we have deduced $M_1 = \dfrac{T_0}{3} \dfrac{(d + r_1)(d - 2r_1)^2}{d^2 - d_1^2}$,

$$M_2 = 5252 \frac{P_1}{R} \frac{(d - 2r_1)^2}{d^2 - d_1^2}.$$

3. Moments Parallel and Perpendicular to the Sections. — These moments above are of fixed direction independent of the angle of the section. This angle varies with the radius of the section. The next step is then obviously, to resolve the above moments parallel and perpendicular to the section. For the ordinary screw whose driving face is a true helicoid this face develops into a straight line, and we will resolve the moments parallel and perpendicular to this line. For sections of varying pitch we will resolve parallel to the tangent at the center of the face. Figure 238 shows an ordinary ogival type of section expanded from its cylindrical shape. Let θ denote the pitch angle, or the angle which the face line makes with a plane perpendicular to the shaft. Then $OB = M_1 =$ fore and aft moment, and $OA = M_2 =$ transverse moment. If M_c denote the resultant moment perpendicular to the face, we have from Fig. 238,

$$M_c = OC + OD = M_1 \cos \theta + M_2 \sin \theta.$$

Similarly, if M_l denote the moment parallel to the face, we have

$$M_l = BD - AC = M_1 \sin \theta - M_2 \cos \theta.$$

Now θ, the pitch angle, depends upon the pitch and the radius. If p denote the pitch and r_1 the radius, we have

$$\tan \theta = \frac{p}{2 \pi r_1} = \frac{p}{d} \times \frac{d}{2 \pi r_1}.$$

Denote the pitch ratio proper, or $\frac{p}{d}$ by a.

Then $$\tan \theta = \frac{ad}{2 \pi r_1}.$$

Whence $$\sin \theta = \frac{ad}{\sqrt{a^2 d^2 + 4 \pi^2 r_1^2}}; \cos \theta = \frac{2 \pi r_1}{\sqrt{a^2 d^2 + 4 \pi^2 r_1^2}}.$$

We have seen above that $M_c = M_1 \cos \theta + M_2 \sin \theta$.

Substituting their values obtained above for M_1, M_2, $\cos \theta$ and $\sin \theta$ and reducing the results, we obtain

$$M_c = \frac{(d - 2 r_1)^2}{(d^2 - d_1^2) \sqrt{a^2 d^2 + 4 \pi^2 r_1^2}} \left[\frac{2 \pi}{3} T_0 r_1 (d + r_1) + \frac{5252 \, ad P_1}{R} \right].$$

Let us next express r_1 and d_1 as fractions of the diameter d, the main dimension. Write $r_1 = \frac{md}{2}$ and $d_1 = cd$. Upon reducing we have

$$M_c = \frac{(1 - m)^2}{(1 - c^2) \sqrt{a^2 + \pi^2 m^2}} \left[\frac{\pi}{6} T_0 md (2 + m) + \frac{5252 \, a P_1}{R} \right].$$

Proceeding in practically the same manner we obtain

$$M_l = \frac{(1-m)^2}{(1-c^2)\sqrt{a^2+\pi^2 m^2}}\left[\frac{T_0}{6}ad(2+m) - 5252\frac{P_1}{R}m\pi\right].$$

In any particular case of design we will know P_1 and R, but generally will not know T_0. The equation connecting T_0 and P_1 is

$$\frac{T_0 p(1-s)R}{33000} = eP_1.$$

Since $p = ad$, this gives

$$T_0 = \frac{33000\, eP_1}{adR(1-s)}.$$

Now in practical cases e approximates but is generally somewhat less than $1-s$ for practical slips, being greater than $1-s$ only for very high slips. So if we assume $e = 1-s$, the result will be to make the value of T_0 generally greater than the truth. In other words, we shall generally introduce a moderate error on the safe side and simplify our expressions enormously. So write $T_0 = \frac{33000\, P_1}{adR}$.

Also introduce the factor 12 in the expressions for M_c and M_l, so that these moments, heretofore expressed in pound-feet, will be expressed in inch-pound units. This is desirable because it is convenient to measure dimensions of the propeller sections in inches. Then substituting in the expression for M_c and M_l, $T_0 = \frac{33000\, P_1}{adR}$, and multiplying by 12 we have after reduction

$$M_c = 63{,}024\frac{(1-m)^2}{(1-c^2)\sqrt{a^2+\pi^2 m^2}}\left[3.29\frac{m(2+m)}{a}+a\right]\frac{P_1}{R},$$

$$M_l = 132{,}000\frac{(1-m)^2}{(1-c^2)\sqrt{a^2+\pi^2 m^2}}[1-m]\frac{P_1}{R}.$$

In the above expressions $\frac{P_1}{R}$ is the factor depending upon the work done. The complicated fractions involve m the fraction of the radius; a, the extreme pitch ratio; and c, the ratio between diameter of propeller and diameter of hub. Hence these complicated fractions can be calculated and plotted once for all. So write

$$C = 63{,}024 \frac{(1-m)^2}{(1-c^2)\sqrt{a^2+\pi^2 m^2}} \left[3.29 \frac{m(2+m)}{a} + a \right],$$

$$L = 132{,}000 \frac{(1-m^2)}{(1-c^2)\sqrt{a^2+\pi^2 m^2}} (1-m).$$

Then finally, $M_c = C \dfrac{P_1}{R}$, $\quad M_l = L \dfrac{P_1}{R}$.

Figure 239 shows curves of C and L plotted upon m for various values of a. For these curves c was taken uniformly at $\frac{2}{8}$. Even if the hub has a different diameter, this is generally an amply close approximation for practical purposes. Since, however, for very large hubs a correction may be needed, there is given in Fig. 240 a "Curve of Correction Factors" for hub diameters. It is seen that for $m = \frac{2}{8}$ the factor is unity. For smaller hubs the factor is less than unity, and for larger hubs greater than unity. Unless, however, the hub diameter is one-third of the propeller diameter or more, it is not worth while to undertake to correct the regular values of C and L in Fig. 239, namely, those for the hub diameter $\frac{2}{8}$ the propeller diameter.

4. **Resisting Moments of Section.** — The above expressions enable us, by the use of Fig. 239, to obtain very readily with sufficient approximation the longitudinal and transverse bending moments at any section of a given propeller of known power and revolutions. It is next in order to consider the resistance of the section, using, as already stated, the developed section. Referring to Fig. 241, let AB, the length of a section in inches, be denoted by l, and CD, its thickness at the center in inches, be denoted by t. The center of gravity will be found on CD at a point G, say. Denote DG by gt, where g is a coefficient. Let I_c, or the moment of inertia about a horizontal axis through G, be denoted by $k_c l t^3$ and I_l or the moment of inertia about CD, by $k_l l^3 t$. Then for the type of section above we have due to M_c:

$$\text{Tension at } A \text{ and } B = \frac{gt}{k_c l t^3} M_c = \frac{g}{k_c} \frac{M_c}{l t^2}.$$

$$\text{Compression at } C = \frac{(1-g)t}{k_c l t^3} M_c = \frac{1-g}{k_c} \frac{M_c}{l t^2}.$$

Due to M_l we have, if B is leading edge,

$$\text{Tension at } A = \text{compression at } B = \frac{\frac{1}{2}l}{k_l l^2 t} M_l = \frac{1}{2 k_l} \frac{M_l}{l^2 t}.$$

These are general expressions. The coefficients g, k_c and k_l depend upon the type of section, and l and t are the dimensions. It will be well, then, to consider the range of values of the coefficients g, k_c and k_l for various possible types of section. The most usual type of section is the ogival, where AB is a straight line and the curve ABC the arc of a circle. This type of section, however, is difficult to reduce to rule, the coefficients varying with the proportions. The ogival section, however, is practically the same as a section with a parabolic back, so the latter may be considered.

In addition to the parabolic back as representing the ordinary type of blade section we will consider two other types of blade of the same maximum thickness. In one the parabola is replaced by a curve of sines. In the other, thickness is equally distributed between face and back, each being a curve of sines. Figure 242 shows the three types of blade section, and below each are given the equation characterizing it, the expressions giving the area in terms of length and thickness, and the value of the coefficients for it. Then for the three types of blade section we have with sufficient approximation:

	No. 1.	No. 2.	No. 3.
Maximum tension at..........	A	A	D
Expression for maximum tension............	$\frac{g}{k_c} \frac{M_c}{lt^2} + \frac{1}{2 k_l} \frac{M_l}{l^2 t}$	$\frac{g}{k_c} \frac{M_c}{lt^2} + \frac{1}{2 k_l} \frac{M_l}{l^2 t}$	$\frac{g}{k_c} \frac{M_c}{l^2 t}$
Maximum compression at.....	C	C	C
Expression for maximum compression............	$\frac{1-g}{k_c} \frac{M_c}{lt^2}$	$\frac{1-g}{k_c} \frac{M_c}{lt^2}$	$\frac{1-g}{k_c} \frac{M_c}{lt^2}$

5. Compressive Stresses. — It is not obvious whether for types 1 and 2 the maximum tension is greater or less than the maximum compression. It is found, however, upon investigation of blades as they are found in practice, that the compression stress is the greatest and the only one that need be considered in the case of material that is as strong in tension as in compression.

It would seem, then, to be the best plan in practice to design the blade thickness from considerations of compression and then determine tension of the blade thus designed. In the rare cases where the tension is found too high it is easy to make the necessary changes. The formula for compression at the center of the back of the blade in pounds per square inch is,

$$\text{Compression} = \frac{1-g}{k_c} \frac{M_c}{lt^2}.$$

Now $M_c = C \dfrac{P_1}{R}$, and it is seen from Fig. 242 that for all these types of blade a safe value for $\dfrac{1-g}{k_c}$ is 14. Then our final formula is, Maximum Compression at Center of Back in Pounds per Square Inch, $= 14 C \dfrac{P_1}{R} \times \dfrac{1}{lt^2}$, where C is obtained from Fig. 239.

We are now in a position to investigate the stress, not only at a root section, but at any point along the radius, by the aid of the above formula and Fig. 239. The result for a blade of rather wide tips and a mean width ratio of .2 is shown in Fig. 243. This shows for various pitch ratios, and plotted on fractions of radius, curves of thickness in center for constant compressive stress, the thickness being expressed always as a fraction of the thickness at .2 the radius. Beyond .2 of the radius these curves are so close together for the various pitch ratios that it is impossible to plot them separately. Below .2 of the radius the curves separate. It is seen that the outer portion of the thickness curve in Fig. 243 is not quite straight, being slightly curved. The curvature is so slight, however, that if we follow the nearly universal practice of making the back of the blade straight radially, the thickness at the tip being not zero but the minimum that can be conveniently cast, the stress per square inch will be practically constant. Unfortunately it is clearly unsafe to make the line of the blade back concave as we go out, thus decreasing thickness and gaining efficiency for high speed propellers. It is true that sometimes the line of the blade back is made concave when the blade has a small hub and is narrow close to the hub, but this is due to a thickening of the inner part of the blade — not a thinning of the outer part. The only practicable method, then, of accomplishing reduction of blade thickness is to use material capable of standing high stress.

Figure 243 indicates that for propellers with small hubs — less than .2 the diameter — the thickness should be determined, not at the root, but at .2 the radius, the straight line of back being extended inward to the hub.

For convenience in design work Fig. 244 has been prepared. This gives values of C, from .2 the radius to .4 the radius for pitch ratios, from .8 to 2.0, thus covering the practical field.

6. **Tensile Stresses.** — Coming back now to the question of tension, it seems that sections of Type 3 are the simplest. The maximum tension for it is the same as the maximum compression. But sections of Type 3 are not desirable for use. For sections of Types 1 and 2 the case is not so simple. Taking the maximum tension as that at A and the maximum compression as that at C, and denoting by t_1 the tension factor or value of maximum tension ÷ maximum compression, we have

$$t_1 = \frac{\frac{g}{k_c}\frac{M_c}{lt^2} + \frac{1}{2k_l}\frac{M_l}{l^2 t}}{1 - \frac{g}{k_c}\frac{M_c}{lt^2}}.$$

Now $\frac{M_l}{M_c} = \frac{L}{C}$, and with sufficient approximation we have from Fig. 242 $g = .4$ and $k_l = .71\ k_c$.

Whence, after simplifying, $t_1 = .666 + 1.17\frac{L}{C}\frac{t}{l}$. L and C are given in Fig. 239, but to facilitate computation Fig. 245 gives curves of $1.17\frac{L}{C}$ from $m = .1$ to $m = .4$, and for final pitch ratios from .6 to 2. This covers the practical ground. For narrow cast-iron blades with solid and hence small hubs it will generally be necessary to determine tensile stress with care.

7. **Stresses Due to Centrifugal Force.** — In addition to the stresses upon a propeller blade due to thrust and torque, there are

stresses due to centrifugal force. These are appreciable. In any given case they can be calculated with sufficient approximation without serious difficulty. If W denote the weight of that portion of a propeller blade outside of the radius r_1 of a given section, r_2 the radius of the center of gravity of the portion of blade and v the circumferential velocity of the center of gravity, while g, as usual, denotes the acceleration due to gravity, then the centrifugal force of the portion of blade may be taken as equivalent to a single force perpendicular to the shaft through the center of gravity of the portion of blade. The amount of the force in pounds will be equal to $\dfrac{W}{g} \dfrac{v^2}{r_2}$. Knowing the force and its line of application, the stresses upon the bounding section of the portion of blade can be determined by well-known methods of applied mechanics.

It appears advisable, however, to make a general mathematical investigation of a case sufficiently simple to admit of such investigation and sufficiently resembling the cases of actual propellers to enable us to apply the results of the mathematical investigation, in a qualitative way at least, to actual propellers. It will be seen that we can thus learn a good deal about the laws governing the stresses of propeller blades caused by centrifugal action.

Fig. 246 shows an elliptical *expanded* blade touching the axis at O. Consider the weight of each section such as CD concentrated at the blade center line at E. Let bd denote the minor axis BN, d being the propeller diameter and b a fraction. The equation of the ellipse referred to O the point where it touches the axis, is

$$y = b \sqrt{2\,dr - 4\,r^2},$$

where r denotes radius and y the semibreadth at radius r.

Then $\qquad$ Breadth $= 2\,y = 2\,b \sqrt{2\,dr - 4\,r^2}$.

Now for r substitute $m\,\dfrac{d}{2}$ where m is fraction of whole radius varying from o at O to 1 at A.

Then $\qquad$ Breadth $= 2\,bd \sqrt{m - m^2}$.

When we come to thickness, the axial thickness is τd, where τ is blade thickness fraction. The tip thickness is not fixed by considerations of strength, being from considerations of castings, etc., usually materially thicker than it need be for strength. We wish in considering centrifugal force to be sure we take the tip thick enough, so will assume it as .15 the axial thickness. It will usually be less in practice for large propellers. Then the back of the blade center being a straight line, the thickness at m is $\tau d(1 - .85m)$.

Assuming the section as parabolic, the area of a section $= \frac{2}{3}$ width $\times$ thickness $= \frac{2}{3} \times 2bd\sqrt{m - m^2} \times \tau d(1 - .85m)$

$$= \frac{4\tau bd^2}{3}(1 - .85m)\sqrt{m - m^2}.$$

We are now able to formulate the elements of curves to be plotted upon m and integrated graphically to obtain the results needed. The element of blade volume = Area of section $\times dr$.

Now $\quad r = \dfrac{md}{2}, \quad dr = \dfrac{d}{2}dm.$

Hence

Element of volume $= \dfrac{2\tau bd^3}{3}(1 - .85m)\sqrt{m - m^2}\,dm.$

Let δ denote weight per cubic foot of the material of the blade

Element of weight $= \dfrac{2\delta\tau bd^3}{3}(1 - .85)m\sqrt{m - m^2}\,dm.$

Element of centrifugal force $= \dfrac{\omega^2 r}{g} \times$ weight $= \dfrac{\omega^2 md}{2g} \times$ weight

$$= \dfrac{\omega^2 \delta\tau bd^4}{3g}m(1 - .85m)\sqrt{m - m^2}\,dm.$$

Let $\quad \displaystyle\int_1^m m(1 - .85m)\sqrt{m - m^2}\,dm = \phi_1(m).$

Then total centrifugal force from the tip to the section m is

$$\dfrac{\omega^2 \delta\tau bd^4}{3g}\phi_1(m).$$

If there is no rake the effect of the centrifugal force is simply to cause a tension over the area. This tension $= \dfrac{\text{force}}{\text{area}}$

$$= \dfrac{\omega^2 \delta\tau bd^4 \phi_1(m)}{3g} \times \dfrac{3}{4\tau bd^2(1 - .85m)\sqrt{m - m^2}}$$

$$= \dfrac{\omega^2 \delta d^2 \phi_1(m)}{4g(1 - .85m)\sqrt{m - m^2}} \text{ in pounds per square foot.}$$

Expressed in pounds per square inch it is $\frac{1}{144}$ of the stress in pounds per square foot. So we have Tension in pounds per square inch due to centrifugal force when there is no rake

$$= \frac{\omega^2 \delta d^2 \phi_1(m)}{576 g (1 - .85 m) \sqrt{m - m^2}} = \frac{\omega^2 \delta d^2}{576 g} \phi_2(m).$$

It appears, then, that for a blade without rake the tensile stress due to centrifugal force varies as the weight per cubic foot of the blade material, as $(\omega d)^2$, or as the square of the tip velocity, and as $\phi_2(m)$ where $\phi_2(m)$ is a quantity depending upon radial position, blade shape, proportions, etc., but independent of size and pitch.

Since it is usually more convenient to express angular velocity by the revolutions per minute, denoted by R, we may substitute $\frac{2\pi R}{60}$ for ω. Also, in order to avoid small decimal factors, multiply numerator and denominator by 1,000,000. Then Tension in pounds per square inch due to centrifugal force when there is no rake

$$= \frac{4 \pi^2 R^2}{3600} \times \frac{\delta d^2}{576 g} \times \frac{1000000}{1000000} \phi_2(m),$$

$$= \frac{\delta d^2 R^2}{1000000} \left[\frac{4000000 \, \pi^2 \phi_2(m)}{3600 \times 576 g} \right],$$

$$= \frac{\delta d^2 R^2}{1000000} \left[\frac{4000000 \, \pi^2}{3600 \times 576 g} \times \frac{\phi_1(m)}{(1 - .85 m) \sqrt{m - m^2}} \right],$$

$$= \frac{\delta d^2 R^2}{1000000} \phi_t.$$

Figure 247 shows curves of $\phi_1(m)$ and ϕ_t. It is seen that $\phi_1(m)$, which is proportional to total centrifugal force, increases always from tip to axis, as might be expected. Since the assumed blade has no area at the axis, ϕ_t, which is proportional to the stress per square inch, is infinity at the axis but falls off very rapidly at first as we go out.

We wish mainly, however, to investigate the effect of rake or inclination upon the stresses on propeller blades due to centrifugal action. Let id denote the total rake of the blade along its center line, where i is a comparatively small fraction, and assume the weight of the section concentrated at the center line. Then idm denotes the rake from the axis to the radius corresponding to m.

Suppose we wish to determine the moment due to centrifugal force about the section corresponding to m_1.

The element of force at m beyond m_1 is, as before,

$$\frac{\omega^2 \delta \tau b d^4}{3 g} m (1 - .85 m) \sqrt{m - m^2}\, dm.$$

Its lever to radius m_1 is $id(m - m_1)$. Hence element of moment

$$= \frac{\omega^2 \delta \tau b i d^5}{3 g} (m - m_1) m (1 - .85 m) \sqrt{m - m^2}\, dm.$$

Moment to $m_1 = \dfrac{\omega^2 \delta \tau b i d^5}{3 g} \left[\int_1^{m_1} m^2 (1 - .85 m) \sqrt{m - m^2}\, dm \right.$

$$\left. - m_1 \int_1^{m_1} m (1 - .85 m) \sqrt{m - m^2}\, dm \right].$$

The second integral is $\phi_1(m)$, but we can denote the whole thing by $\phi_3(m_1)$ and after obtaining results by graphic integration use m instead of m_1. Then we have

$$\text{Moment from tip to } m = \frac{\omega^2 \delta \tau b i d^5}{3 g} \phi_3(m) = M', \text{ say.}$$

Now the moment M_1 is in the plane through the axis and the center line of the blade. Its effect upon the section is best ascertained by resolving it parallel and perpendicular to the section.

If θ be the pitch angle at radius r, $\tan \theta = \dfrac{p}{2\pi r} = \dfrac{p}{\pi m d} = \dfrac{a}{\pi m}$, if we use a to denote the pitch ratio $\dfrac{p}{d}$.

Then $\quad \sin \theta = \dfrac{a}{\sqrt{a^2 + \pi^2 m^2}} \quad \cos \theta = \dfrac{\pi m}{\sqrt{a^2 + \pi^2 m^2}}.$

If M_C' and M_L' denote the moments resolved perpendicular and parallel to the blade face we have

$$M_C' = M' \cos \theta = \frac{\omega^2 \delta \tau b i d^5}{3 g} \pi \phi_3(m) \frac{m}{\sqrt{a^2 + \pi^2 m^2}},$$

$$M_L' = M' \sin \theta = \frac{\omega^2 \delta \tau b i d^5}{3 g} \phi_3(m) \frac{a}{\sqrt{a^2 + \pi^2 m^2}}.$$

Finally, by applying at the center of the section forces equal and opposite to the forces producing the moments, we have the section

PROPULSION

affected by a force and two couples. The force is the same as the outward force when there is no rake. The couples are perpendicular and parallel to the section and their moments are given above. The result of the force and couples is as follows, reference being had to Fig. 241, where B is the leading edge:

1. The force causes a certain tension over the whole section.
2. The perpendicular couple causes compression at C and tension at A and B.
3. The parallel couple causes tension at A and compression at B.

Now from consideration of thrust and torque only we have already found that the maximum compression is at C and the maximum tension at A. Centrifugal action evidently increases the tension at A more than at B. Hence, as regards tension we need consider the action at A only.

As regards compression, when we neglect centrifugal action this is a maximum at C. The tension due to the force decreases compression at B and C equal amounts. Then the parallel moment increases compression at B and the perpendicular moment increases compression at C. We need to find which increase is the greater, and if C has greater compression from centrifugal action we need consider C only.

The necessary coefficients for the parabolic sections are found in Fig. 242. Consider the tension increases at A first. We have three increases:

Due to force alone in pounds per square inch, $\dfrac{\omega^2 \delta d^2}{576\,g}\phi_2(m)$.

Due to perpendicular moment, $\dfrac{8.75 \times 12\,M_C'}{ll^2}$, where the factor 12 has been introduced because we wish stresses per square inch and M_C' was calculated in pound-foot units.

Now in feet $l = 2\,bd\,\sqrt{m - m^2} = 24\,bd\,\sqrt{m - m^2}$ in inches.
Also $t = \tau d\,(1 - .85\,m)$ in feet $= 12\,\tau d\,(1 - .85\,m)$ in inches.
So the tension per square inch at A due to the perpendicular moment is

$$\dfrac{105\,M_C'}{24 \times 144\,b\tau^2 d^3\,\sqrt{m - m^2}\,(1 - .85\,m)^2}.$$

Substituting the value of M_C'

Tension at A due to perpendicular moment

$$= \frac{35\,\pi\omega^2\delta\tau bid^5\phi_3(m) \times m}{1152 \times 3\,gb\tau^2 d^3 \sqrt{m-m^2} \times (1-.85\,m)^2\sqrt{a^2+\pi^2 m^2}}$$

$$= \frac{35}{3456}\frac{\pi}{g}\frac{i}{\tau}\frac{\omega^2\delta d^2 \phi_3(m) \times m}{\sqrt{m-m^2}\,(1-.85\,m)^2\sqrt{a^2+\pi^2 m^2}}.$$

Due to parallel moment

$$\text{Tension at } A = \frac{15 \times 12\,M_L'}{l^2 t}.$$

Reducing this similarly we have

Tension at A due to parallel moment

$$= \frac{5}{576}\frac{1}{g}\omega^2\delta\frac{i}{b}d^2\frac{\delta_3(m)\,a}{(m-m^2)(1-.85\,m)\sqrt{a^2+\pi^2 m^2}}.$$

Suppose now we denote by N the tension per square inch due to centrifugal force only and express these other tensions in terms of N:

We have

$$N = \frac{\omega^2\delta d^2}{576\,g}\phi_2(m) = \frac{\omega^2\delta d^2}{576\,g}\frac{\phi_1(m)}{(1-.85\,m)(m-m^2)^{\frac{1}{2}}}.$$

Then Tension at A due to perpendicular moment

$$= \frac{35\,\pi}{6}N\frac{i}{\tau}\frac{\phi_3(m)}{\phi_1(m)}\frac{m}{(1-.85\,m)\sqrt{a^2+\pi^2 m^2}}$$

$$= \frac{2}{3}N\frac{i}{\tau}\phi_4.$$

Tension at A due to parallel moment

$$= 5N\frac{i}{b}\frac{\phi_3(m)}{\phi_1(m)}\frac{a}{(m-m^2)^{\frac{1}{2}}\sqrt{a^2+\pi^2 m^2}} = N\frac{i}{b}\phi_5.$$

In the above ϕ_4 and ϕ_5 involve a as well as m and should be expressed by contour diagrams.

Consider now the compression at C due to the perpendicular moment. This is $\dfrac{13.125 \times 12 \times M_C'}{lt^2}$. As before in inches

$$l = 24\,bd\,(m-m^2)^{\frac{1}{2}},$$
$$t = 12\,\tau d\,(1-.85\,m).$$

Hence compression

$$= \frac{13.125 \times 12\, Mc'}{24 \times 144 \times b\tau^2 d^3 (m-m^2)^{\frac{1}{2}}(1-.85\,m)^2},$$

$$= \frac{13.125}{288} \times \frac{\pi}{3g}\, \frac{\omega^2 \delta\tau bid^5 \phi_3(m)\, m}{b\tau^2 d^3 (m-m^2)^{\frac{1}{2}}(1-.85\,m)^2 \sqrt{a^2+\pi^2 m^2}},$$

$$= \frac{26.25}{576\,g}\, \frac{\pi}{3}\, \omega^2 \delta d^2 \frac{i}{\tau}\, \frac{\phi_3(m)\, m}{(m-m^2)^{\frac{1}{2}}(1-.85\,m)^2 \sqrt{a^2+\pi^2 m^2}}.$$

And in terms of N

Compression at C

$$= N\frac{i}{\tau}\, \frac{26.25\,\pi}{3}\, \frac{\phi_3(m)}{\phi_1(m)}\, \frac{m}{(1-.85\,m)\sqrt{a^2+\pi^2 m^2}} = N\frac{i}{\tau}\phi_4.$$

We can now express the ratios between extra compression at C and compression at B due to parallel moment. The latter is the same as tension at A due to parallel moment

$$= 5N\frac{i}{b}\, \frac{\phi_3(m)}{\phi_1(m)}\, \frac{a}{(m-m^2)^{\frac{1}{2}}\sqrt{a^2+\pi^2 m^2}}.$$

$$\frac{\text{Extra Compression at } C}{\text{Parallel moment Compression at } B} = \frac{26.25\,\pi}{15}\, \frac{b}{\tau}\, \frac{m}{a}\, \frac{(m-m^2)^{\frac{1}{2}}}{1-.85\,m}.$$

Now we may safely say that in practice b is greater than 4τ. If we put $b = 4\tau$, $\pi = \tfrac{22}{7}$, we have for above ratio,

$$22\, \frac{m}{a}\, \frac{(m-m^2)^{\frac{1}{2}}}{1-.85\,m}.$$

The hub is such that m may be taken as .2 or more. Putting $m = .2$ we have ratio above $= \dfrac{2.12}{a}$. So for propellers in practice the extra compression at C due to centrifugal action will always be greater than that at B due to the parallel moment. When, too, we recollect that there is a large opposing tension at B due to the perpendicular moment, it is obvious that the maximum compression is at C, and only that need be considered.

Figures 248 and 249 show contours of ϕ_4 and ϕ_5 plotted on a and m and curves of $\phi_1(m)$, ϕ_t and $\phi_3(m)$ which involve m only are shown in Fig. 247.

We have finally for stresses due to centrifugal forces

Tension in pounds per square inch neglecting rake = $N = \dfrac{\delta d^2 R^2}{1000000} \phi_t.$

Extra compression at center of blade back = $N\left(\dfrac{i}{\tau}\phi_4 - 1\right)$

$= \dfrac{\delta d^2 R^2}{1000000} \phi_t \left(\dfrac{i}{\tau}\phi_4 - 1\right)$ in pounds per square inch.

Extra tension at following edge = $N\left(\dfrac{2}{3}\dfrac{i}{\tau}\phi_4 + \dfrac{i}{b}\phi_5 + 1\right)$

$= \dfrac{\delta d^2 R^2}{1000000} \phi_t \left(\dfrac{2}{3}\dfrac{i}{\tau}\phi_4 + \dfrac{i}{b}\phi_5 + \right)$ in pounds per square inch.

In the above i is ratio between rake and diameter, τ is ratio between axial thickness and diameter, and b is ratio between maximum blade width and diameter and may be taken as 1.188 (mean width ratio).

The above formulæ and the accompanying figures apply strictly only to blades whose expansion is an ellipse touching the axis and whose tip thickness is .15 the axial thickness.

The methods used can be followed to determine $\phi_1(m)$, $\phi_t \phi_3(m)$, ϕ_4 and ϕ_5 for blades of any type, but the results of Figs. 247, 248 and 249 can be applied in practice with sufficient approximation to any oval blade that does not depart widely from the elliptical form.

Since centrifugal stresses increase as the square of the tip speed, they evidently need to be given much more careful consideration for quick running propellers than for those of moderate speed. Thus, suppose we had a manganese bronze propeller for which $dR = 4000$, or the tip speed is over 12,000 feet per minute. For manganese bronze $\delta = 525$ about. Then $N = 525 \times 16 \phi_t = 8400 \phi_t$. For $m = .3$, $\phi_t = .135$ about, so $N = 1134$. If the pitch ratio is about unity, $\phi_4 = 2\frac{1}{3}$ about, and if $\dfrac{i}{\tau}$ has the value of 3 or the rake is three times the axial thickness, $\dfrac{i}{\tau}\phi_4 - 1 = 6$, or increase in compressive stress at .3 the radius is the large amount of 6700 lbs. This is an extreme but not impossible case. As tip speed falls off, stresses due to centrifugal force decrease rapidly, but it would seem the part of wisdom to avoid them entirely by avoiding backward rake. Moreover, it seems advisable when tip speed is very high to give a moderate forward or negative rake, thus opposing the tensile and compressive stresses due to the work done by opposite stresses due to the centrifugal forces. When backing, centrifugal force would add to the natural stresses, but propellers are not worked backward at maximum speed.

In calculating stresses due to centrifugal force we need values of δ or weight per cubic foot of the various materials used for propeller blades. For manganese bronze or composition we may use 525 for δ, for cast iron 450 and for cast steel 475.

8. Stresses Allowable in Practice. — While for quick-running propellers centrifugal stresses must be calculated separately, in the majority of cases they are not very serious and may be allowed for by using a low stress in our main strength formulæ.

Compressive stress in lbs. per sq. in. $= S_c = 14 \, C \dfrac{P_1}{R} \times \dfrac{1}{lt^2}.$

Tensile stress in lbs. per sq. in. $= S_T = S_c (.666 + 1.17) \dfrac{L}{C} \dfrac{t}{l}.$

In applying these formulæ to the root section of any blade we will know C, P_1, R and l. Then we fix t by giving S_c a suitable value and calculate S_T to see if that has a suitable value. Now what are suitable values of S_c for the various materials of which we make propeller blades? They cannot be fixed arbitrarily from consideration of only the tensile and compressive strengths of the material. For one thing our formulæ are approximations only. In order to apply the methods of Applied Mechanics we start by developing the cylindrical section of the blade into an ideal plane section. It is probable that this ideal section is materially weaker than the actual section, especially in the case of propellers of varying pitch. Hence, if this were the only perturbing factor, we could allow high stresses in the formulæ, because the stresses per formulæ would be greater than the true stresses. But when we consider the conditions of operation of propellers we find other very serious perturbing factors which we cannot reduce to rule. In the formula, P_1 is the average power absorbed by the blade. But even in still water the blade, owing to inequalities of wake, will absorb more power than the average at one portion of the revolution and less at another. And in disturbed water, what with the motion of the water and the pitching of the ship, the blade is liable to encounter stresses very much in excess of those due the average power which it absorbs. This is especially likely to be true of turbine driven propellers. With reciprocating engines, when a propeller encounters abnormal resistance the engine will soon slow down, the kinetic energy of the moving parts being rapidly absorbed. With turbines, however, we are likely to have the kinetic energy of the moving parts per square foot of disc area much greater than for reciprocating engines, and the flywheel action, so to speak, of the moving parts is then capable of causing a relatively greater extra stress.

To determine with scientific accuracy allowable stresses for use in the formula we would probably have to test to destruction full-sized propellers — which is impracticable. The next best thing is to find from the formula the stresses shown by actual propellers which have been successful in service, and also those of propellers which have shown weakness in service. We can thus establish, with sufficient accuracy for practical purposes, the maximum stresses that can be tolerated. The advantage in this connection of a formula upon a sound theoretical basis is that a stress found satisfactory for a fine-pitched, quick-running propeller, for instance, will be almost equally satisfactory for a coarse-pitched propeller, and vice versa, so that satisfactory allowable stresses can be deduced from less data than would be necessary for a formula partaking largely of the rule of thumb nature.

There are advantages in the use of a simple semi-graphic method which will enable data from completed vessels to be recorded for use in design work.

We have deduced as the final formula for S_c the compressive stress in pounds per square inch for blades of the usual ogival section

$$S_c = 14\, C \frac{P_1}{R} \frac{1}{lt^2}$$

where C is a coefficient depending on radius and pitch ratio, P_1 is the power absorbed by the blade, R denotes revolutions per minute of the propeller and l and t are width and thickness respectively of the blade in inches. Also we should generally use in determining S_c the values of C, l and t at about .2 the radius of the propeller. Let us now express l and t in terms of coefficients and ratios already used.

Put $l = 12\,chd$ where d is diameter in feet, h is mean width ratio and c is a coefficient depending upon the shape of the blade.

It is not such a simple matter to determine a rigorous expression for t, because the tip thickness is more or less independent of the root thickness.

If τd denote axial thickness as usual, and $k\tau d$ the tip thickness, we have for .2 the radius

$$t = 12\,\tau d\,[k + .8\,(1 - k)] = 12\,\tau d\,(.8 + .2\,k).$$

In practice k is seldom much less than .1 or greater than .2 Now $k = 0$, $t = 9.6\,\tau d$, $k = .1$, $t = 9.84\,\tau d$, $k = .2$, $t = 10.08\,\tau d$. So it is a sufficient approximation for practical purposes to put

$$t = 10\,\tau d.$$

So, returning to the stress formula, we have

$$S_c = 14\,C\,\frac{P_1}{R} \times \frac{1}{12\,chd} \times \frac{1}{100\,\tau^2 d^2} = \frac{14\,C}{1200}\,\frac{P_1}{Rd^3}\,\frac{1}{ch\tau^2}.$$

Let $C_1 = \dfrac{14\,C}{1200}$. Figure 250 shows plotted upon pitch ratio a curve of C_1 for .2 the radius.

Then
$$S_c = \frac{C_1 P_1}{Rd^3} \times \frac{1}{ch\tau^2}.$$

Suppose, now, we put $\dfrac{C_1 P_1}{Rd^3} = x$, $\quad ch\tau^2 = y$:

then we have
$$S_c = \frac{x}{y}.$$

Figure 251 shows contours of values of S_c plotted on x and y. In the case of a given propeller we know or can readily calculate $ch\tau^2$ and $\dfrac{C_1 P_1}{Rd^3}$. Hence, we can locate a spot on Fig. 251 corresponding to the propeller which will show the root compression or value of S_c in pounds per square inch. Figure 251 shows by crosses a number of spots each of which corresponds to an actual propeller. They are nearly all for vessels of war, and all for manganese bronze or other strong alloy. It is desirable, when using the method for design work, to reproduce Fig. 251 on a large scale. It is evident from Fig. 251 that the designers of the propellers referred to differed widely as to the allowable stress. No. 11 refers to a destroyer which would very seldom develop maximum power, and then only in smooth water. But even for such vessels it is not advisable to go to such stresses. No. 14 was a vessel which much exceeded her designed power, on trial, and also sprung her propeller blades. With manganese bronze and similar alloys now available it is inadvisable to exceed 15,000 lbs. even for destroyers. For other fast men-of-war which seldom develop full power, suitable stresses, based upon full power, are 10,000 to 12,000 pounds per square inch. For merchant vessels, always at nearly full speed, particularly passenger steamers that are driven hard in rough weather, it is not advisable to exceed 5000 to 6000 lbs. The above all refer to blades of manganese bronze and similar alloys. Good cast-steel propellers can be given the same stresses as those of manganese bronze.

For cast iron it is advisable not to exceed 5000 lbs. for compression and 2000 lbs. for tension.

As already stated, designers differ widely as to the proper stresses to allow for propeller blades. It is a simple matter for any designer with an accumulation of data for actual propellers to record it on a large diagram similar to Fig. 251 and form his own conclusions as to the stresses which he will allow in a particular case.

While it is desirable for a designer fully to understand all details involved in determining propeller blade thickness, it may be pointed out that when centrifugal forces are not serious, and the blade thickness is to be fixed from considerations of compressive stress only, Figs. 250 and 251 are all that need be consulted. For when number of blades, diameter and pitch have been determined we can determine P_1, R and d. C_1 can be taken from Fig. 250, so we will know $\dfrac{C_1 P_1}{R d^3}$. From the blade outline we can determine h and c, the latter usually falling between .6 and .8 in practice. Thus in a practical case, after having calculated ch we need only to determine τ.

So we enter Fig. 251 with the value of $\frac{C_1 P_1}{R d^3}$, and from the stress chosen determine $ch\tau^2$, and ch being known τ^2 and τ are readily determined.

9. Connections of Detachable Blades. — While somewhat apart from the question of strength of propeller blades it seems advisable to consider briefly the question of the strength of the connections of detachable blades. We have seen that the formulæ for transverse and fore and aft moments in pound-feet are:

$$\text{Fore and aft moment } M_1 = \frac{T_0}{3} \frac{(d + R_1)(d - 2 r_1)^2}{d^2 - d_1^2}.$$

$$\text{Transverse moment } M_2 = 5252 \frac{P_1}{R} \frac{(d - 2 r_1)^2}{d^2 - d_1^2}.$$

Also with a margin for safety we may write $T_0 = \frac{33000 \, P_1}{a d R}$.

Making this substitution and multiplying by 12 to reduce moments to inch-pounds, we have:

$$M_1 = 132{,}000 \frac{P_1}{R} \frac{(d + r_1)(d - 2 r_1)^2}{a d (d^2 - d_1^2)},$$

$$M_2 = 63{,}024 \frac{P_1}{R} \frac{(d - 2 r_1)^2}{d^2 - d_1^2}.$$

Now with sufficient approximation we may write $d_1 = \frac{2}{8} d$.
Also we may take r_1 or the radius to hub flange to which the blade is bolted, as $\frac{1}{8}$ the propeller radius with a slight error on the safe side. Substituting and reducing, we have in round numbers

$$M_1 = \frac{116000}{a} \frac{P_1}{R}, \qquad M_2 = 52{,}400 \frac{P_1}{R}.$$

These two moments may be compounded into a single moment whose direction makes with the direction of the shaft axis $\tan^{-1} \frac{524 \, a}{1160}$ and whose amount in inch-pound units is

$$100 \frac{P_1}{R} \sqrt{(524)^2 + \left(\frac{1160}{a}\right)^2} = H \frac{P_1}{R} \text{ say.}$$

The amount and angle of the moment depend upon $\frac{P_1}{R}$ and a only.

Figure 252 shows plotted upon a, or extreme pitch ratio, curves of values of the angle of inclination of the moment and of the coefficient H.

The moment above must be resisted by the bolts securing the blade flange to the hub and the flange itself. The bolts are, of course, disposed on each side of the direction of the moment, and it is good practice to use more bolts for the side where the bolts are in tension when going ahead. Thus, if there are nine bolts in all, five will be in tension when going ahead and four in tension when backing.

Theoretically, the blade flange will pivot under stress about some point on its extreme circumference and the leverage of each bolt will be the length of a perpendicular from its center to a line drawn through the pivoting point tangent to the circumference.

For a conventional assumption, however, which is an adequate approximation, we may take the effective leverage of each bolt in tension as the diameter of the circle through the center of the bolts.

Investigation of actual propellers upon this basis indicates 3000 pounds per square inch as a fair average of the stresses allowed on steel flange bolts by designers, the actual stresses varying from less than 2000 pounds to some 4000 pounds.

Even after making all allowances for the conditions of service it would seem that 3000 pounds per square inch is a low stress for such bolts and that 4000 pounds or more might be used without apprehension.

For quick running propellers the stress taken account of should include that due to centrifugal force upon the blade. The expression for force in pounds is

$$\frac{\omega^2 \delta \tau b d^4}{3g} \phi_1(m),$$

and for moment in pound-feet,

$$\frac{\omega^2 \delta \tau b i d^5}{3g} \phi_4(m).$$

The moment may be taken as parallel to the shaft axis. It is seen from Fig. 247 that we may, with fair approximation, use .09 for

$\phi_1(m)$ and .04 for $\phi_3(m)$. Substituting these values, and putting $g = 32.16$ and $\omega = \dfrac{2\pi R}{60}$, we have:

$$\text{Force in pounds} = \frac{\delta \tau b d^4 R^2}{97{,}755}.$$

$$\text{Moment in pound-feet} = \frac{\delta \tau b i d^5 R^2}{219{,}950}.$$

31. Design of Propellers

1. Number and Location. — Nearly all the matters of detail involved in propeller design have been already considered, but it is proposed briefly to review the general considerations involved, and illustrate the methods already explained by working out a few examples. The question of the number and location of propellers is not very often an open one at any stage of the design, being usually fixed by practical or other considerations which have little to do directly with propeller efficiency. From the point of view of propeller efficiency only, the best location for a propeller is in the center line, as far aft as possible. In the center line it gets the maximum benefit from the wake and the farther aft it is the less the thrust deduction. Practical considerations of protection from damage require the screw to be forward of the rudder, but a suitable arrangement by which the screw was located abaft the rudder, so that its suction would not produce appreciable thrust deduction, would undoubtedly increase efficiency of propulsion. Since, however, suction will have no retarding effect upon a fore and aft plane, about the most that can be done in practice to reduce thrust deduction upon a single screw vessel is to make the after portion as fine as possible. In many cases there might be more done in this direction than is done. Fineness at the water surface is what is needed.

As to vertical location, it is the usual practice to locate screws as low as possible. For seagoing ships this is desirable to reduce racing, and even for ships intended for smooth water service only, it is generally necessary, because such vessels are usually of shallow draft, and to get the propeller sufficiently beneath the water surface it must be placed low. But propellers are not placed so low that their tips project below the keel if this can be avoided.

This is simply to reduce risk of damage in case of grounding, and in some cases it is necessary to ignore this risk and allow the propeller tips to go below the keel.

There is little doubt, that contrary to what is generally supposed, a propeller for smooth water work is more efficient the closer it is to the surface, provided it is not so close that it draws air from the surface. This, for the reason that in this position it gets the greatest useful reaction from the wake. Frictional, wave, and stream line wakes are all strongest near the surface.

One is apt to conceive of the frictional wake as a vertical belt of nearly uniform horizontal thickness. But an examination of Figs. 50 to 59, and careful observations of actual ships, would seem to indicate that the frictional wake abreast the stern widens rapidly as we approach the surface, and in fact we may almost regard the wake as made up of a vertical layer close to the ship and a horizontal layer extending out some distance from the ship, but not extending deeply into the water. The higher a center line propeller is the more it gains from the vertical layer, and if it is high enough to reach the horizontal layer it gains still more. But as already pointed out, it is necessary to give a good submergence to the screw of a seagoing vessel to avoid racing in a seaway. A broken shaft is too serious a matter to be risked in order to secure slightly greater propulsive efficiency in smooth water. Furthermore, in rough water a deeply submerged screw which does not race will have much higher propulsive efficiency than one close to the surface that is racing constantly. So in practice we usually find screws of seagoing vessels immersed as deeply as practicable.

The best location for a side propeller is probably the nearest location practicable to the best location for a center line propeller. Where twin screws are fitted they would, under this rule, be placed as far aft as possible and as close to the center line as possible.

It must be said, however, that the fore and aft location of a side screw appears to have surprisingly little effect upon its efficiency. We saw in considering actual and virtual shaft deviations that for a four-screw vessel the after pair were about as badly off in this respect as the forward pair. We would expect, however, *a priori*, that a side screw well forward would usually have greater virtual shaft deviation than one well aft, and would also gain less from the wake and have a greater thrust deduction.

It is undesirable to place screws so that their tips are too close to the surface of the hull. When a screw tip strikes the belt of eddying water adjacent to the hull, the virtual blows resulting are communicated to the ship, shaking rivets loose and causing vibration. The irregular forces upon the propellers also cause vibration of the ship.

In some twin-screw ships this trouble has been partially avoided by leaving an opening in the dead wood abreast the propellers. This saves the ship, and with large propellers of moderate speed of revolution the tips can be brought quite close to one another without giving trouble. For small, quick-turning propellers, such as those fitted with turbines, vibrations are very likely to be set up unless the blade tips are kept well clear of the hull, say 30 inches to 36 inches. It seems a pity to lose any of the beneficial action of the wake, and it is possible that if the hull abreast the propeller tip were made of circular shape, with the shaft as a center, specially strengthened to stand the pounding, and the propeller tips fitted close to the hull so that they caught the dead water through a large arc, the beneficial effect of the wake might be had without very objectionable vibration, though such propellers would probably be noisy. That is a matter, however, which could be determined only by a full-sized trial. The only solution now known to be successful is to keep the blade tips well clear and accept the slightly reduced efficiency.

When triple screws are fitted, it is obviously desirable that the races from the side screws should almost or entirely clear the disc of the center screw. This result is best attained when the side screws are forward of and above the center screw.

For a side screw located well forward the question of virtual deviation due to the water rising up and closing in aft is frequently given less attention than it should receive, resulting in loss of efficiency and vibration from the screws.

When four screws are fitted the after pair are located in the natural location of twin screws, and the forward pair are placed forward and higher so as to avoid interference as far as possible. These forward screws, if badly placed, are liable to serious virtual shaft deviations, and the questions of their location, shaft angles, etc., should receive most careful consideration. They may, from their high location, get a better reaction from the wake, and hence not lose in propulsive efficiency as compared with the after screws.

The number of screws depends upon various considerations. If there is no limit to diameter and revolutions, there is no question that the single screw should be the most efficient. There is probably not much to choose between twin and triple screws as regards propulsive efficiency. Quadruple screws are likely to be somewhat the least efficient as regards location. In practice, however, in a given case, diameter and revolutions are not unrestricted, and the number of screws is apt to be fixed from other considerations than those of slight differences of efficiency due to number of screws.

Twin screws were adopted for men-of-war primarily to secure greater immunity from complete breakdown, greater protection of screws and engines on account of smaller size, and ability to do some maneuvering independent of the rudder. The same considerations influenced the adoption of twin screws for high-class passenger vessels, but another consideration came in here. With the very great powers used for such vessels the engines or shafts became too large with single screws. This consideration has also largely influenced the adoption of triple and quadruple screws.

With the advent of the turbine the question of revolutions — already of importance in fixing the number of screws for quick-running engines — became a very important one.

For steam economy and weight saving the turbine should use high revolutions. But a propeller which absorbs great power at high revolutions must be given so much diameter in proportion to its pitch that its efficiency becomes too small. Hence, with turbines we usually find three or four shafts. In the early days of turbines multiple screws were often fitted — two or three on each shaft. This practice has now been abandoned, however, as a result of experience, the present practice being to fit but one screw on each shaft.

While in many cases with turbines it is desirable for the best economy to use three screws, it is rather difficult with three screws to secure satisfactory arrangements for the rudder post and rudder.

Still it is possible to do this, and three screws are used until questions of economy or size of units drive us to the use of four screws.

2. Direction of Rotation. — Obviously, when we have a center line screw it will give the same efficiency whether it is right-handed or left-handed. Hence the direction of rotation of single screws and of the center screw of triple screws is immaterial. The desirable direction of rotation of side screws depends upon considerations of water flow and shaft obliquity already discussed in detail.

For ships as they are, in the vast majority of cases, it seems probable that side screws would be slightly more efficient if outward turning. For side screws very far aft, with shafts supported by struts, so that the fittings for carrying the shafts do not interfere with the natural water flow, it matters little as regards efficiency whether the screws be in or out turning. With shaft webs approaching the horizontal, the side screws should be outturning for efficiency. With shaft webs approaching the vertical, they would be more efficient if inturning. Such shaft webs are, however, practically unknown. Side screws materially forward of the stern, however their shafts are supported, should turn outward for the best efficiency.

As regards efficiency, then, in about all practical cases side screws should be outturning. For maneuvering by means of the screws alone, when a vessel has not steerage way, outturning screws are distinctly preferable for practically all types of vessesl. For many vessels this consideration alone would outweigh minor difference of efficiency, but as outturning screws have the advantage as regards efficiency in nearly all practical cases, they should be adopted in the vast majority of cases. Cases may occur where it is a matter of indifference, and cases are conceivable where, as with vertical shaft webs, inturning screws are more efficient, but outturning screws should be the rule and inturning screws should be fitted only for good and sufficient reasons, which in practice will exist very seldom indeed.

3. Number of Blades. — When the number and location of propellers are settled and it becomes necessary to get out finally the design of the propeller, we will know the power which it is expected to absorb and the revolutions it is to make. The speed of the ship will be known, and we can estimate the wake factor and thus determine the speed of advance. About the first thing to be settled is the number of blades. Two-bladed propellers are hardly worth considering for jobs of any size. Figure 217 indicates that appreciable gain in efficiency is not to be expected from them, and they are distinctly inferior as regards uniformity of turning moment and vibration.

So, in practice, the choice will lie between three blades and four blades. Model experiments of a comparative nature appear to indicate that three-bladed propellers are essentially more efficient than four-bladed.

It is seen from Fig. 216, however, which probably exaggerates, if anything, the inferiority of four-bladed propellers that this inferiority is small, and it may well happen in practice that a four-bladed propeller exactly adapted to the conditions will be superior to a three-bladed propeller not so well designed.

Many designers are firm believers in the superiority of the four-bladed screw as well as many sea-going engineers. Probably in rough water the four-bladed screw will show a slightly more uniform turning moment and less tendency to produce vibration. But some of the fastest Atlantic liners that are driven at top speed in fair weather and foul have three-bladed screws. All things considered, there are probably few cases in practice where with equally good design the three-bladed propeller is not somewhat to be preferred. It should always be lighter and cheaper, and this is a matter worthy of consideration, especially when the propeller is to be made of an expensive composition.

In some large four-screw turbine jobs, two of the screws have been made four-bladed and two three-bladed with satisfactory results.

With this combination the chance of objectionable vibration due to synchronism is practically eliminated. Where special reasons such as this exist, or where strong prejudices exist, it may be advisable to use four-bladed propellers, but in the vast majority of cases three blades should be used.

We have seen in Section 25 that propellers with solid hubs are slightly more efficient than those with detachable blades. The difference is small, however, except for quick-running propellers, which are usually of small diameter. There are great difficulties in the way of accurately casting and finishing large propellers with solid hubs — say propellers over 12 feet in diameter. Hence, such propellers should nearly always be made with detachable blades.

4. Material of Blades. — For the material of propeller blades we have a choice between cast iron, cast steel, and some copper alloy, such as composition, manganese bronze or other special alloy. Forged steel blades have been used, but are not found now.

For such a vessel as a tugboat, with its wheel near the surface and liable to strike floating objects, cast iron is regarded as desirable. Its brittleness and weakness here become virtues, for when a blade strikes something it breaks without endangering the shaft or engine, and it is cheaper and shorter to renew the propeller than the shaft or portions of the engine. Cast steel is superior to cast iron in strength and is largely used for merchant work.

Manganese bronze and other special alloys can now be had with strength equal or superior to that of cast steel. They can be given a better surface, and from the point of view of efficiency of propulsion are decidedly the better materials. They have two drawbacks. The first cost is higher, and through galvanic action they are liable to cause excessive corrosion of the portion of the ship's structure adjacent to them. This damage can, however, be neutralized in practice by the use of zinc plates properly secured to the hull.

A very serious objection to iron and steel blades is their tendency to corrode. The backs of the blades where there is eddying water probably mixed with air seem peculiarly subject to extensive and rapid corrosion.

The practical conclusion is that noncorrosive blades should by all means be used, unless their first cost prohibits them for the job in hand or unless for special reasons cast iron is indicated.

But in many cases cast iron or steel blades as a gift would be in the end more expensive than noncorrosive blades, owing to the loss of efficiency and greater coal consumption caused by their extra friction when corroded. This extra friction is the more objectionable the finer the pitch of the propeller.

5. Width of Blades. — The blade area of a propeller of given diameter and pitch varies directly as the mean width ratio. While it has sometimes been thought that comparatively small changes of blade area had large effects upon propeller action and efficiency, this view is hardly sustained by practical experience. When cavitation is not present, rather large changes in blade area produce quite small effects. It should be remembered, too, that in practice change of blade area involves change of blade section with attendant change of virtual pitch.

The $\rho\delta$ diagrams indicate clearly that when cavitation is absent the best mean width ratio is between .25 and .30. For mean width ratio of .35 the efficiency is appreciably reduced, and for wider blades still it falls off quite rapidly. These conclusions are for very smooth blades. In practice blades become more or less roughened and foul, and when this is the case the wider blades will have the greater loss of efficiency.

The conclusion indicated as a practical rule is that where cavitation is not to be feared the best all-round mean width ratio is about .25 or less. To avoid cavitation wider blades up to a mean width ratio of .35 or so should be used, even with thin blades of hollow-backed type. In extreme cases even wider blades may be required, in spite of their excessive friction loss.

6. Examples of Design. — The principles governing propeller design and the application of the methods that have been given will now be illustrated by some typical cases.

First Case. — Design the propeller for a turbine Atlantic liner which develops 80,000 shaft horse-power upon four screws making 200 revolutions per minute each and has a speed of 28 knots. Here we may take the propeller power as 20,000. The first thing necessary is to estimate the wake factor. In the case of a job of such importance this would be done nowadays from model experiments. Let us suppose that we are considering the after screws and that the wake factor is 10 per cent.

Then $\quad V_A = .9 \times 28 = 25.2.$

So $\quad \rho = \dfrac{200 \sqrt{20,000}}{(25.2)^{2.5}} = 8.87.$

Also $\quad d = \delta \dfrac{(20,000 \times 25.2)^{\frac{1}{4}}}{(200)^{\frac{1}{2}}} = .2608\ \delta.$

We are now prepared to enter the $\rho\delta$ diagrams (Figs. 211 to 214).

PROPULSION

Since, however, we know that this is a case where cavitation is to be carefully provided against, we would expect to use a blade of wide type, so we will use only Fig. 214 for a mean width ratio of .35. In Fig. 214 for $\rho = 8.87$ the best pitch ratio is 1.140 and the best value of $\delta = 57.4$. Then diameter $d = .2608 \times 57.4 = 14'.97$ and pitch $= 14.97 \times 1.140 = 17'.07$, the real slip being 25.2 per cent. These for a blade thickness fraction of .03. Now the power P_1 absorbed by each blade is 6667. From Fig. 250 for a pitch ratio of 1.14, $C_1 = 910$ and $(14.97)^3 = 3355$.

Hence for Fig. 251 $\quad x = \dfrac{C_1 P_1}{Rd^3} = \dfrac{910 \times 6667}{200 \times 3355} = 9.04.$

Now it seems advisable in such a job to keep the stress down to moderate limits. So let us try for it 7500 lbs. per square inch. From Fig. 251 where $x = 9.04$ and compressive stress is 7500, $y = ch\tau^2 = .0012$ about. Now we know $h = .35$, and if $c = \tfrac{2}{3}$, which will be somewhere near the truth, we have $\tau^2 = \dfrac{.0012 \times 3}{.35 \times 2} = .00514$; $\tau = .072$, axial thickness $= 12''.9$. Now Fig. 214 being based upon a blade thickness fraction of .03, it is necessary to correct the results obtained by using Fig. 215. From this figure when $\rho = 8.87$ for each .01 increase of τ the diameter should be decreased 1.1 per cent and the pitch ratio increased 0.9 per cent. So the total decrease in diameter would be $1.1 \times 4.2 = 4.62$ per cent and increase of pitch $0.9 \times 4.2 = 3.78$ per cent. This would make the diameter $14.97 \times .9538 = 14'.28$, pitch $17.07 \times 1.0378 = 17'.72$. If we allowed a stress of 10,000 lbs. per sq. in. which might be admissible in such a high-class job as this we would have from Fig. 251 $y = ch\tau^2 = .0009$. Whence, for $c = \tfrac{2}{3}$ $h = .35$,

$$\tau^2 = \dfrac{.0009 \times 3}{.35 \times 2} = .00386.$$

$\tau = .0621$, axial thickness $= 11''.2$.

The reduction of thickness is not very much, but we could probably stand an axial thickness of 12 inches.

Now the tip speed will be over 9000 feet per minute and even with the best possible shape of blade section some cavitation is to be expected. So as much increase of slip would involve rapid falling off of efficiency, it would seem advisable to make the propeller a little large in order to provide against this and adopt as the final dimensions: Diameter 15 feet, pitch 17 feet 6 inches, mean width ratio .35, axial blade thickness 12 inches. The propeller efficiency to be expected, barring cavitation, is about 67 per cent.

Second Case. — Design the propeller for a large twin-screw turbine destroyer to make 34 knots with 25,000 shaft horse-power at 800 revolutions per minute, the wake fraction being .03.

Then $V_A = 34 \times .97 = 32.98$,

$$\rho = \frac{800 \sqrt{12,500}}{(32.98)^{2.5}} = 14.3,$$

$$d = \frac{\delta (12,500 \times 32.98)^{\frac{1}{2}}}{(800)^{\frac{3}{2}}} = .1001\ \delta.$$

This too is a case where cavitation is to be carefully guarded against, so we consider only Fig. 214.

From this figure for $\rho = 14.3$ the best pitch ratio is 1.004 and $\delta = 60.3$, the propeller efficiency being about 62 per cent.

Then $d = 6'.036, \quad p = 6'.06.$

Consider now blade thickness, $P_1 = 4167$, and from Fig. 250

$$C_1 = 1015, \quad \text{also } d^3 = 220.$$

Then from Fig. 251

$$x = \frac{C_1 P_1}{R d^3} = \frac{1015 \times 4167}{800 \times 220} = 24.$$

This is a value of x beyond the limits of Fig. 251, but to use this method a designer should prepare an enlarged and extended copy of Fig. 251. In this case we wish to use a high stress, say 12,000 lbs. It will be found that using this stress in an enlarged copy of Fig. 251 we have for $x = 24, ch\tau^2 = .0020$.

In this case, too, we may put $c = \frac{2}{3}$ and we have $h = .35$. Then

$$\tau^2 = \frac{.0020 \times 3}{2 \times .35} = .00857, \quad \tau = .0926.$$

Axial thickness $= 6\frac{3}{4}''$.

In this case, too, there would be a decrease of diameter of about 7 per cent and an increase of pitch of nearly 6 per cent from Fig. 215. But with a tip speed of about 15,000 feet per minute there will almost certainly be cavitation, and it is not safe to reduce the diameter. It does seem advisable, however, to increase the pitch slightly to provide against excessive slip. So the dimensions indicated are: Diameter 6 feet $\frac{1}{2}$ inch, pitch 6 feet 5 inches, mean width ratio .35, axial blade thickness $6\frac{3}{4}$ inches. The propeller efficiency to be expected in the absence of cavitation is about 62 per cent, but this is a case where the actual efficiency depends largely upon the amount of cavitation. Some cavitation is almost unavoidable. The propeller in this case would be cast with solid hub. We thus lose the possibility of varying the pitch and hence adjusting the propeller to the engines after trial. In cases where there is uncertainty it is possible virtually to provide for this, however, by making the propeller originally a little large. If trials show it too large, blade tip can be cut off to suit, being careful not to throw the propeller out of balance.

Third Case — Design the propeller for a twin-screw gunboat to make 17 knots with 3700 I.H.P. at 156 revolutions per minute, the wake fraction being .08.

Then $V_A = 17 \times .92 = 15.64$. We are dealing now with I.H.P. and must estimate the propeller power. Assume it .9 of the I.H.P. Then

$$P = \frac{3700 \times .9}{2} = 1665,$$

$$\rho = \frac{165 \sqrt{1665}}{(15.64)^{2.5}} = 6.96, \quad d = \frac{\delta \, (1665 \times 15.64)^{\frac{1}{2}}}{(165)^{\frac{3}{2}}} = .181 \, \delta.$$

This is a case where with proper blade section we need not seriously apprehend cavitation. Hence we should try all four $\rho\delta$ diagrams. The results are tabulated below:

$\rho\delta$ diagrams, Fig. No............	211	212	213	214
Mean width ratio................	.20	.25	.30	.35
Blade thickness fractions, (Standard)................	.06	.05	.04	.03
Maximum standard efficiency.....	.700	.706	.702	.687
Best pitch ratio.................	1.222	1.208	1.200	1.233
Best δ	54.6	55.5	56.2	55.6
Diameter — d....................	9'.88	10'.05	10'.17	10'.06
Pitch — p.....................	12'.07	12'.14	12'.20	12'.41
d^2...........................	964	1015	1052	1018
P_1 (power absorbed by each blade)	555	555	555	555
C_1 from Fig. 250................	873	878	880	870
Value of $x = \dfrac{C_1 P_1}{R d^4}$	3.05	2.91	2.81	2.87
Assumed nominal stress 10,000 lbs.				
Value of $ch\tau^2$ from Fig. 251......	.0003	.0003	.0003	.0003
Value of ch, assuming $c = \tfrac{2}{3}$.....	.1333	.1667	.2	.2333
Value of τ^2	.00225	.0018	.0015	.001286
Value of τ.....................	.0474	.0424	.0387	.0359
Departure of τ from standard.....	−.0126	−.0076	−.0013	+.0059
Per cent change of diameter from Fig. 215....................	+1.58	+0.95	+0.16	−0.74
Per cent change of pitch ratio from Fig. 215....................	−1.49	−0.04	−0.14	+0.65
New pitch ratio.................	1.204	1.197	1.198	1.241
New diameter..................	10'.04	10'.15	10'.19	9'.98
New pitch.....................	12.09	12.15	12.21	12.39

This is a case where we have a wide range of choice of width with little change of efficiency. It is evident, too, from the $\rho\delta$ diagrams that we may change diameter and pitch through a range of 10 per cent each without much effect upon efficiency.

Where cavitation is not to be feared the best all-round mean width ratio is about .25, and using this we would finally adopt: Diameter 10 feet 2 inches, pitch 12 feet 2 inches, M.W.R. .25, B.T.F. .0424, axial thickness of blade, 6 inches.

Fourth Case. — Design the propeller for a large single screw cargo vessel to make 12 knots with 4000 I.H.P. at 78 R.P.M., the wake fraction being .26.

For this case $V_1 = 12 \times .74 = 8.88$, and if we assume the propeller power to be .9 the I.H.P. we have $P = 3600$.
Then

$$\rho = \frac{78 \sqrt{3600}}{(8.8)^{2.5}} = 19.9, \quad d = \frac{\delta(3600 \times 8.88)^{\frac{1}{2}}}{78^{\frac{3}{2}}} = .309\,\delta.$$

This being a case of moderate tip speed we will consider mean width ratios of .2 and .25 only. Results are tabulated below:

$p\partial$ diagram, Fig. No.	211	212
Mean width ratio	.20	.25
Blade thickness fraction, (Standard)	.06	.05
Maximum efficiency for value of ρ	.602	.600
Best pitch ratio	.860	.801
Best ∂	62.8	63.0
Diameter	19'.41	19'.47
Pitch	16'.70	17'.35
Real slip, per cent	30.9	33.5
Apparent slip, per cent	6.7	10.1
d^2	7313	7381
P_1 — power absorbed by each blade	1200	1200
C_1 — from Fig. 250	1200	1150
Value of $x\left(\dfrac{C_1 P_1}{d^2 R}\right)$	2.52	2.39
From Fig. 251 for cast steel, — $ch\tau^2 =$ nominal stress 5000	.00050	.00048
If $c = \frac{3}{4}$, $ch =$	.1333	.1667
τ^2	.00375	.00288
τ	.061	.054
The departure of τ from standard values is too small to take account of. If we wish to use a 4-bladed screw we have from Fig. 216 — Diameter co-efficient for $\rho = 19.9$	.946	.946
Pitch coefficient for $\rho = 19.9$	.971	.971
Diameter of 4-bladed screw	18.36	18.42
Pitch of 4-bladed screw	16.21	16.85
The pitch ratio being low, suppose we assume a pitch ratio of 1.1. Then for a 3-bladed screw we have efficiency	.590	.585
∂	58.0	58.1
Diameter	17'.92	17'.95
Pitch	19'.71	19'.94
For corresponding 4-bladed screw — diameter	16'.95	16'.98
For corresponding 4-bladed screw — Pitch	19'.14	19'.17

The above example is a very interesting one and illustrates several facts apt to be lost sight of.

In the first place, the large vessel of low speed as built has a value of ρ entirely too large for good propeller efficiency. The ρ value is materially larger than that of the destroyer case and the maximum possible propeller efficiency less. It is true that the large cargo vessel should approach the maximum, while the destroyer is apt to lose through cavitation. In spite of the low propeller efficiency the efficiency of propulsion of the cargo vessel may be good. Such vessels are apt to have a hull efficiency greater than unity, which brings up their efficiency of propulsion, but the fact remains that their efficiency of propulsion would be better still if they could be given more efficient propellers. In order to do this the ρ value must be less. Now ρ can be reduced by reducing the revolutions, but this will result in increased diameter, which is already by no means small. Also reduced revolutions are almost certain to be objectionable as regards the engine. Another practicable method of reducing p is to use twin screws, but this has obvious objections. The trouble is essentially the same as encountered with moderate speed turbine vessels, namely, that the desirable engine revolutions are too high for a propeller of high efficiency. There is a further trouble, namely, that the propeller of high efficiency may require an impossibly large diameter. Still, the best solution of the problem is the same as for the turbine, namely, a satisfactory speed reduction gear of high efficiency, so that both engine and propeller can be given the revolutions best suited to their needs.

It will be observed that the propeller of best efficiency has to work at a very high real slip. This essential condition is masked in practice by the fact that the wake fraction is large, so that the apparent slip is very much below the real slip. In fact, for such vessels very good results may be obtained when the apparent slip is zero.

The fact that the best we can do in such cases is to work a propeller of fine pitch, and hence low maximum efficiency, at a high slip, so that its efficiency is well below its maximum, is the main reason for the rapid reduction of efficiency with large values of p. For small values of p propellers can usually be worked much closer to their point of maximum efficiency.

It will be observed that while for the .25 M.W.R. the best pitch ratio is .891, this can be made 1.1 with a reduction of possible efficiency from .600 to .585 only. But the diameter can be reduced thus from $19'.47$ to $17'.95$ or over $18''$, the pitch rising from $17'.35$ to $19'.74$. If a four-bladed screw is used the diameter can be reduced still more.

32. Paddle Propulsion

The vast majority of sea-going vessels are propelled by screws, and vessels using paddle wheels are practically all engaged in channel, bay, lake or river service.

1. **General Features.** — It is obvious that a paddle wheel through its construction and method of operation approaches more nearly than the screw propeller the ideal frictionless propelling apparatus

discussed in Section 22. If, for instance, we regard a paddle wheel as discharging directly astern a column of water of area equal to the area of a paddle float and with velocity equal to the difference between the peripheral velocity of the center of the float and the speed of advance of the ship, and make the further assumption that the action is frictionless and that the water is discharged without change of pressure we have an ideal propelling instrument to which Fig. 171 applies.

This leads us to the conclusion that if A denote the area of a paddle float, V the speed of advance in knots and P the shaft horse-power absorbed by the paddle wheel $\frac{P}{AV^2} = \phi(e) = \phi(s) = K'$, where the coefficient K' is a function of the slip. For paddle wheels the slip is generally reckoned with reference to the peripheral speed of the paddle centers. If V_p denote the peripheral speed of the paddle centers in knots and V the speed of advance of the vessel in knots,
$$s = \frac{V_p - V}{V_p}.$$

2. Fixed Blades. — The earliest paddle wheels had the blades on radical lines, as indicated diagrammatically in Fig. 253, and many paddle wheels are still of this type.

Figures 254 and 255 trace out the successive positions of a single float with reference to still water for 30 per cent slip and 10 per cent slip respectively. The direction and relative amounts of the velocities of the inner and outer edges of the floats are also indicated.

The line marked $W.L.$ indicates a water line such that the blade has its upper edge immersed in its deepest position about one half of its breadth. There is of course minimum obliquity of action when the blade is vertical, in its deepest position, and it is desirable that the blade should do as much work as possible when deeply immersed. That would require it to enter the water edgewise, or nearly so. It is evident from Figs. 254 and 255 that radial blades will not be moving edgewise with respect to still water at the time they reach the water surface. This result may, of course, be accomplished by setting the blades at suitable fixed angles. But fixed blades so set are usually regarded as undesirable, perhaps without good reason.

In the United States the development of wheels which will not suffer from excessive obliquity of blades at entering and leaving has been toward wheels of large diameter and wide narrow floats of small immersion. This line of development was facilitated by the type of engine usually fitted on paddle steamers.

Furthermore, broadly speaking, paddle steamers in the United States have been for service in smooth waters, and hence could be designed for a small immersion of floats which would be inadvisable in rough water service.

3. **Feathering Blades.** — In Great Britain, influenced perhaps originally by the fact that many of the finest and fastest paddle steamers were for service across the English Channel and had to be prepared to encounter rough weather, paddle wheels are almost universally fitted with feathering blades.

As indicated diagrammatically in Fig. 256, a blade is pivoted about its center, the pivots being carried by the framing of the wheel proper, which revolves about A. Each blade has an arm perpendicular to it on its back, to which is attached a link, and the other end of the link is connected to a center K eccentric from A. The point K is very simply determined. The positions of H, G and F are obviously fixed by the positions desired for a blade entering the water, leaving the water and at maximum submergence. Then K is the center of the circle passing through H, G and F.

It is very common in practice to fit feathering paddle blades as indicated in Fig. 256, where the planes of the entering and leaving blades intersect the circle of blade centers vertically above the shaft. Paddle wheels have been fitted where the blades remained vertical throughout the revolution, but this is not done now.

It might seem very simple from Figs. 254 to 255 to determine the proper angles for blades entering and leaving the water, but the actual problem is one of extreme complexity. Figs. 254 and 255 show velocities with reference to water at rest, and this is far from the conditions of practical operation.

The water upon which a paddle wheel acts has been previously disturbed by the ship, the amount of disturbance varying with the speed. Moreover, each paddle enters water which has been disturbed by the preceding paddles. There is little question that in practically all cases of side paddle wheels the paddles enter water which has already a sternward motion. Stream line action and the action of preceding paddles will both give the water a sternward motion, and even if the wheel is located at a wave crest — as is desirable — the forward motion due to the wave motion will be less than the other two.

For stern wheels stream line and wave action will give the water a forward motion, the action of preceding paddles a rearward motion, and it is not possible without extensive experiments to lay down any general conclusions.

4. Comparison of Fixed and Feathering Blades. — Paddle wheels with feathering blades are heavier, more complicated and more expensive than wheels of the same size with fixed blades. But in practice they can be made materially smaller in diameter for the same efficiency, and also can be given greater depth of immersion — resulting in a larger virtual area of paddle for a given actual size. This is an important consideration for high-speed paddle vessels. The smaller the wheel the higher the engine revolutions, and it is usually desirable as regards weight and space to increase the revolutions of paddle boat engines when directly connected. In practice fast high-powered paddle boats are usually fitted with feathering blades, fixed blades being used when the revolutions are low and the diameter of wheel great, or for service in remote rivers where simplicity is essential.

5. Paddle Wheel Location. — While it is not proposed to consider structural details, some considerations affecting paddle wheel design will now be taken up. In practice, paddle wheel vessels are side wheelers or stern wheelers. In side wheelers the wheels are located somewhere near the center of length. It is advisable to locate them so that they work in a crest of the transverse waves caused by the ship, or at any rate not in a hollow. When working in a crest there is a virtual wave wake favoring efficiency, while in a hollow the wave wake is prejudicial to efficiency. The stream line wake in which side wheels work is prejudicial to efficiency, so that side paddle wheels usually have a virtual negative wake. Also the wash from the wheels increases the frictional resistance of the rear of the ship and produces a virtual thrust deduction.

Side wheels cannot be placed very far forward or aft of the center of ships of ordinary form without danger of under or over immersion through changes of trim, incident to service.

Stern wheel boats are of the wide flat type and the draft aft does not vary much in service.

Stern wheels are so located that the wake due to stream line and wave action is in their favor, and they will cause but little thrust deduction as a rule, so that, broadly speaking, the stern wheel may be expected to be more efficient as an instrument of propulsion than side wheels.

It is very desirable to fix the heights of all paddle wheels so that the desired immersion will be had when the vessel is under way. This can readily be done by model basin experiments in advance, and for the best results with feathering wheels the question of blade angles at entrance in and departure from the water should also be investigated experimentally.

The immersion of paddles is varied somewhat with the service. For seagoing boats the immersion of the upper edge of the paddle in its lowest position is seldom less than $\frac{1}{2}$ its breadth and as great as .8 its breadth. For smooth water service the immersion is usually less, $\frac{1}{8}$ to $\frac{1}{4}$ the breadth. The desirable immersion depends somewhat upon the type of float. A very long narrow float on a large wheel may have its upper edge immersed its whole breadth without loss of efficiency.

6. Dimensions and Proportions of Paddle Wheels. — One of the most important questions arising in the design of any type of paddle wheel is the determination of the dimensions of the blades, buckets or floats, as they are variously designated.

These are sometimes curved, but seldom curved much, and may be taken as rectangular. The length or horizontal dimension of the float is always greater than its width or radial dimension. There is found in practice a difference in proportions between feathering and fixed floats. For feathering floats the length is usually about 3 times the width, though shorter floats have often been fitted. For fixed floats the length is seldom less than 4 times, and may be in extreme cases 7 or 8 times the width. This difference of practice naturally arises from the fact that floats are usually made as long as possible from practical considerations, as tending to efficiency, and then as wide as necessary to absorb the power. For side wheels, floats are, however, seldom longer than $\frac{1}{2}$ the beam even for vessels always in smooth water, and for seagoing vessels it is not regarded as good practice to make them longer than about $\frac{1}{3}$ the beam.

The float area is dependent primarily upon the power absorbed and the slip. We have seen that the theoretical formula involved is $\frac{P}{AV^3} = K'$. This may be rewritten $A = K\frac{I}{V^3}$ where A is area of two floats (one on each side) in square feet, I is indicated horsepower and proportional in a given case to P, V is speed of ship in knots and K is a coefficient depending primarily upon the slip and secondarily upon a large number of minor factors, such as wake, thrust deduction, float proportions, number and immersion, etc.

Hence K may be expected to vary a great deal from ship to ship, but fortunately it is not necessary to know it with minute accuracy.

Analysis of a number of published trial results for paddle steamers, nearly all with feathering floats, appears to indicate that a reasonable expression for the average value of K will be, for slips used in practice ranging say from .10 to .30,

$$K = 212.5 - 375\, s.$$

From the nature of the case individual values of K may be expected to vary materially from the average. A long narrow blade deeply immersed may be expected to show a much smaller value of K than a short wide blade with its upper edge barely immersed.

Then a suitable paddle area may be determined approximately by the formula $A = (212.5 - 375\, s)\frac{I}{V^3}$. It must be remembered that in the above A is total area in square feet of two paddles when side wheels are fitted, and s is slip based upon the peripheral velocity of the centers of paddles.

It is desirable to keep the slips of paddle wheels low. For feathering floats .15 is frequently aimed at, and for fixed floats .20. Knowing the speed of the ship and the desired slip, the peripheral velocity of the mean diameter of the paddle wheel upon which slip is based is known, and this in conjunction with the desired engine revolutions fixes the mean diameter of the wheel.

The desired float area being known, the float dimensions are determined, enabling all dimensions to the wheel to be fixed. If these are found suitable the desired blade angles at entry and departure will govern the details of gear for feathering blades when such are fitted.

As regards number of blades it is a very common practice with fixed blades to fit one for each foot of outside diameter of wheel. This number should not be exceeded for wheels of good size and may be reduced by 20 per cent or so without detriment. The spacing of feathering blades is greater than that of fixed blades, partly because such blades are usually relatively deeper than fixed blades and partly because of the additional complications of feathering gear for blades close together.

With feathering blades there are sometimes fitted one for each foot of radius but a greater number are usually regarded as desirable, say about 3 blades to each 2 feet of radius.

33. Jet Propulsion

1. **General Considerations.** — Jet propulsion has never been used except experimentally. In jet propulsion water is taken into a ship, where it passes through some form of pump or impelling apparatus and then delivered astern through suitable pipes. Many schemes for jet propulsion have been brought forward in the past, usually including methods for diverting the jets sidewise as desired, in order to gain maneuvering power.

While some schemes of jet propulsion have been actually tried, none has proved so efficient as the screw propeller or paddle wheel. Hence, jet propulsion is of academic interest only and will not be given detailed consideration.

That any system of jet propulsion involving any form of impelling apparatus known at present must be inefficient will be evident from Fig. 171. It will be found from this that, even with frictionless impelling apparatus, if there is not to be a great loss through slip the pipes to get the water into and out of the ship must be so large that they will involve very serious increase in skin friction to say nothing of eddy losses. If pipes are made small there is unavoidably a great loss by slip, and still larger loss by friction in the pipes. Furthermore, any pump or impelling apparatus now known is not materially more efficient in communicating velocity to a given quantity of water than the screw propeller or the paddle wheel.

Hence, jet propulsion, involving taking water in large amount into the ship and discharging it again, is with any known form of impelling apparatus necessarily less efficient than the screw and the paddle, which operate in the water outside the ship.

Since the essential inefficiency of jet propulsion as a method of utilizing the power of ordinary engines has become evident, some inventors have attempted to devise apparatus specially adapted to jet propulsion in which power is developed more economically than in engines driving propellers and paddle wheels. Efforts along this line have not hitherto been successful.

CHAPTER IV

Trials and their Analysis

34. Measured Courses

1. Features Desirable for Measured Miles. — Trials for the determination of speed must be made over a course of known length, unless by trials already made over such a course the relation between revolutions of the propellers and speed through the water has been established so that a speed trial may be conducted in free route. The measured course may be long or short. The difficulties of locating, measuring and marking a satisfactory long course are evidently much greater than for a short course, and nearly all accurately measured and marked courses are one nautical mile long. For a number of years, however, four-hour full-speed trials of United States naval vessels were held on long deep water courses extending to the northward of Cape Ann on the Massachusetts coast. The length used was carefully determined in each case so that the vessel would run about two hours in each direction and four or five vessels or more were anchored on the course for the double purpose of defining it and of making observations of the tidal current during trials. Of late years, however, four-hour full-speed trials have been made in free route by the standardized screw method. For standardizing the screw or determining the relation between speed and revolutions, trials are usually held on a course one measured mile in length near Rockland, Me. This course is shown in Fig. 257.

It is seen that the course is defined by four range buoys, one at each end of the measured mile and one a mile from each end. These buoys, however, are for steering purposes only. The ends of the course are fixed by ranges established on shore, each with a front and rear signal or beacon. When these signals are in line the observer is at one end of the course, which, as shown, is perpendicular to the range lines.

The desirable features for a measured mile course in tidal waters are enumerated below.

If they were all present in any particular case the course would be ideal. In practice it is necessary to be satisfied with a reasonable approximation to the ideal.

1. The range marks on shore at each end of the course should be well separated — say $\frac{2}{3}$ the length of the course or more — and should by the transit of the front signal past the back signal mark definitely and sharply the instant of crossing the range. This is best attained when both front and back signals show against the sky.

2. The situation should be such that the course is not far from shore and fairly well protected, insuring smooth water when the local wind conditions are favorable.

3. There should be plenty of room at each end of the course for turning.

4. The course should be so situated that the ship making runs over it need never cross or obstruct a channel or fairway that is much used.

5. The tidal current should be small and always parallel to the course.

6. The depth of water should be sufficient, so that the resistance of the ship using the course is practically the same as in deep water.

As regards most of the features enumerated above, the Rockland course, shown in Fig. 257, approximates fairly closely to the ideal. It has the disadvantage of being rather remote from most of the building yards whose vessels must use it.

It would be better if the front and back signals marking the ranges were further separated and showed above the sky line. It may be noted in this connection that if the range marks do not show against the sky a course running north and south is not so good as one running east and west. If the ranges are to the west of the course the marks are difficult to pick up in the afternoon, and if they are to the east they are difficult to pick up in the forenoon.

35. Conduct of Speed and Power Trials

1. General Considerations. — Vessels may be given many kinds of trials, as of speed and power, of fuel economy, maneuvering capacity, etc. We need consider the first named only.

Speed and power trials may be considered from the point of view of (*a*) the owner, (*b*) the designer, or (*c*) the builder. In some cases, as for vessels of war built in government establishments, the owner, designer and builder are one; frequently for vessels of war the owner and designer are one; and usually for merchant ships, and sometimes for vessels of war, the designer and builder are one.

From whatever point of view we consider speed trials, however, they are primarily of importance for new and untried vessels. For such vessels the owner wishes to know what his ship will do in service and from the results of progressive speed and power trials he can generally closely estimate the results to be expected in service. The designer wishes to know what the ship actually does under known trial conditions in order that he may utilize the information in preparing subsequent designs. The builder is generally required to guarantee certain results to be demonstrated by trial before the ship leaves his hands and at times wishes to develop on trial certain results not exacted by his contract, but which may be of use to him in a business way. Apart from this he is apt to consider that trials conducted at his expense should be reduced to the lowest terms.

As a result of various conflicting considerations the most that can usually be expected for speed and power trials of a new ship in the builder's hands is the determination of corresponding values of speed, revolutions, and power over a reasonable range from the maximum down, at one displacement and under favorable conditions of wind and weather. Such a trial is usually called a progressive speed trial and appears to have been first developed in Great Britain by Mr. William Denny. Concerning this development Mr. William Froude said in a paper before the Institution of Naval Architects, April 7, 1876:

" Mr. Denny has taken the bold but well-considered step of discarding the conventional type of measured mile trials which, as regards the speeds tried, have long been limited to full speed and half boiler power. Mr. Denny now tries each of his ships at four or even at five speeds; and the result is that he obtains fair data for a complete curve of indicated horse-power from the lowest to the highest speeds; whereas with trials on the ordinary system we obtain merely two spots in the curve, and these at comparatively high speeds, the intermediate or lower portion of the curve being left uninvestigated."

2. **Accuracy Possible in Progressive Trial Results.** — The determination of accurate results on a progressive trial is by no means the simple matter it might seem at first. Approximate results are, of course, readily obtained, but for the results of progressive trials to be of real value for the designer they should be quite accurate. What we need are simultaneous values of speed of the vessel, power indicated by the machinery and revolutions per minute of the engines, determined for a sufficient number of speeds covering a good range to enable accurate curves of power and revolutions as ordinates to be drawn on speeds as abscissae throughout the range covered by the trials.

If we had available a measured course in perfectly still, calm, deep water, and wished to determine the most reliable curves from a definite number of runs, it would evidently be desirable to run back and forth, increasing and decreasing the speed or revolutions by equal amounts between successive runs. Observing on each run the time and revolutions on the course and taking indicator cards for the power determination, we could plot curves through points obtained by the observations.

Progressive trials are not made on ideal courses, as above. Even if they were, it would seldom happen that the data obtained would be absolutely consistent and harmonious. It is probable that on a course in still water the time on the course would be determined with a good deal of accuracy. But even with a long straight run at each end before coming on the course — an important point frequently neglected — the speed on the course is seldom absolutely uniform. Unless steam is actually blowing off all the time the boiler pressure is always going up or down — it may be very slowly with skilled firing, it may be with sufficient rapidity to cause quite an appreciable change in speed while on the course. Moreover, the rudder is constantly being used more or less, and even when put over to a small angle only it has a noticeable effect upon the speed. This is a matter of practical importance in the conduct of trials which does not always receive proper attention.

Then the indicator — even the best — is not an instrument of precision. If several sets of cards are taken during a run the powers worked out from them will differ materially. Professor Peabody, an authority on indicators, considers that even " under favorable circumstances the unavoidable error of a steam engine indicator is likely to be from two to five per cent."

If the indicated horse-power is determined on the measured course, not less than three sets of cards should be obtained and the average of all good cards used in determining the average power. At times some cards are obviously defective, and these should be thrown out.

For single-screw ships the revolutions and speed vary together, and there are no serious complications from the inevitable slight variations in revolutions, except that sometimes there is doubt as to the proper revolutions to use with the indicator cards for the determination of power. But with twin-screw ships the revolutions and power of the two engines are not identical on any run. The only thing that can be done in such cases is to try to have the port and starboard revolutions during each run as nearly the same as possible and use the average of the two results. With two screws, unless the propellers differ more than they should, we may safely assume that at a given speed and the same revolutions, each engine will require the same power. In practice, owing to minor differences in propellers, and differences in engine friction, the assumption is not exact. But it is near enough, and is, in fact, the only one we can make.

With three screws, however, the case is different. At full speed, with everything wide open, the central engine will differ from the side engines as regards both power and revolutions, even if identical in size with them. When it comes to the runs at reduced speed, we may for a given speed have enormous variations in the power distribution. It would seem proper in such cases, where the engines are identical, to be careful to have the steam pressure in the H.P.

valve chests and the linking up the same for all three engines on each run. Otherwise the curves of slip of the center and side screws will be very erratic. With four screws the case is even more complicated.

For such vessels, where each engine is independent, it may be necessary to plot results upon speed — plotting separate curves of revolutions for each engine. But even here equally good results can be obtained by plotting results upon the average revolutions of one pair of engines — plotting, upon these revolutions, a curve of the average revolutions of the other engine or pair of engines. For turbine installations, where the turbines are in tandem, the steam passing from one turbine to another, this method is distinctly preferable.

When we come to turbines we meet the difficulty of determining the actual power exerted by them. Several methods are used — all based upon the fact that the twist of the shafting is proportional to the torque of the turbine. This twist is a small quantity in any case, and its accurate determination experimentally is difficult. It is probable, however, that as the use of turbines extends the accuracy of their power determination will be improved. With an accurate torsion meter the determination of shaft horse-power will be much simpler and easier than the determination of indicated horse-power by means of indicators.

3. Elimination of Tidal Current Effects. — It is evident from what has been said that even on an imaginary still-water course a progressive trial would not be free from doubts and difficulties in connection with obtaining and plotting the results.

Actual measured courses, however, must be laid off in a tideway where tidal currents varying in direction and magnitude are encountered. No course is suitable for a progressive trial unless the tidal current is practically parallel to the course. Slight cross currents are nearly always present, however. When they are present the steering on the course should always be by compass and not by buoys or other fixed fore and aft ranges. By always steering a compass course parallel to the true range the effect of slight cross currents is eliminated. So we will consider from now on only the current parallel to the course. Suppose, first, that the current is constant and that we make two runs at the same true speed — one with and one against the current.

Suppose V is the true constant speed of the two runs, C the constant but unknown speed of current and V_1, V_2, the apparent speeds of the successive runs. Then $V_1 = V + C$, $V_2 = V - C$ whence $V = \frac{1}{2}(V_1 + V_2)$, or the true speed through the water is the average of the two apparent speeds with and against the current. Sometimes the true speed is taken as that corresponding to the average time of the runs with and against the current. This is incorrect. The true speed for two runs with and against a constant current being the average of the two apparent speeds, it is a common practice to make the runs of a progressive trial in pairs — one run being made in each direction at the same speed. There are two objections to this. One is that the tidal current changes between runs. The other — often more serious in practice — arises from the fact that in practice the successive runs are made not at the same speed but at different speeds, and the average horse-power is not the proper horse-power for the average speed. Figure 260 illustrates this, in an exaggerated form. A and B are points on a curve of horse-power plotted on speed corresponding to two runs. C is the point on the curve corresponding to the average speed, while D, midway of the straight line joining A and B, is the average horse-power.

The first source of error, the change of tidal current, can be largely, but not entirely, eliminated by making a series of runs over the course at one speed and obtaining the true speed from the apparent speeds by the method of successive means. This is illustrated below with four runs — the apparent speeds being V_1, V_2, V_3, V_4.

Apparent Speeds.	First Means.	Second Means.	Final Means.
V_1	$\frac{V_1 + V_2}{2}$	$\frac{V_1 + 2V_2 + V_3}{4}$	$\frac{V_1 + 3V_2 + 3V_3 + V_4}{8}$
V_2	$\frac{V_2 + V_3}{2}$	$\frac{V_2 + 2V_3 + V_4}{4}$	
V_3	$\frac{V_3 + V_4}{2}$		
V_4			

The first means are simply the averages of the successive pairs of runs. The second means are the averages of the successive pairs of first means, and so on. There appears to be a difference of opinion as to whether, when there are more than four runs, the true speed should be taken as the final mean, or the average of the second means. As appears above, for four runs the two are the same.

Now if n denote the number of a run of a series we can always express C, the strength of the current, in the form

$$C = a + bn + cn^2 + dn^3 + en^4 + \ldots,$$

using as many terms as there are runs in the series. Suppose, for instance, there are four runs. Then we have

$$C = a + bn + cn^2 + dn^3.$$

Denote by C_1, C_2, C_3, C_4 the actual current strength of the four successive runs.

Then
$$C_1 = a + b + c + d,$$
$$C_2 = a + 2b + 4c + 8d,$$
$$C_3 = a + 3b + 9c + 27d,$$
$$C_4 = a + 4b + 16c + 64d.$$

These are four equations from which we could determine the four unknown quantities a, b, c, d. Hence, no matter what the current strength of the successive runs, we could always find values of the coefficients a, b, c and d such that we can represent the current by

$$C = a + bn + cn^2 + dn^3.$$

On solving the equations above for a, b, c, d we have

$$a = \tfrac{1}{6}(24 C_1 - 36 C_2 + 24 C_3 - 6 C_4),$$
$$b = \tfrac{1}{6}(-26 C_1 + 57 C_2 - 42 C_3 + 11 C_4),$$
$$c = \tfrac{1}{6}(9 C_1 - 24 C_2 + 21 C_3 - 6 C_4),$$
$$d = \tfrac{1}{6}(- C_1 + 3 C_2 - 3 C_3 + C_4).$$

Now consider further the final mean result. We have, if V denotes the true constant speed of the four runs,

$$V_1 = V + C_1 = V + a + b + c + d,$$
$$V_2 = V - C_2 = V - a - 2b - 4c - 8d,$$
$$V_3 = V + C_3 = V + a + 3b + 9c + 27d,$$
$$V_4 = V - C_4 = V - a - 4b - 16c - 64d.$$

Final mean = $\frac{1}{4}(V_1 + 3V_2 + 3V_3 + V_4)$. Upon substituting for V_1, V_2, etc., in this expression, their values above in terms of V and the coefficients a, b, c and d, we finally have, after reduction,

Final mean = $V - \frac{3}{4}d = V - \frac{1}{8}(-C_1 + 3C_2 - 3C_3 + C_4)$.

In case only three runs are made the current formula is

$$C = a + bn + cn^2,$$

and the currents of the successive runs are C_1, C_2 and C_3. For this case

Final mean = $\frac{1}{4}(V_1 + 2V_2 + V_3) = V + \frac{1}{2}c = V + \frac{1}{4}(C_1 - 2C_2 + C_3)$.

Then the final mean is not the true speed unless the rate of change of the tidal current and the timing of the runs is such that for four runs

$$-(C_1 - C_4) + 3(C_2 - C_3) = 0,$$

and for three runs

$$C_1 + C_3 = 2C_2.$$

This will happen exactly only by accident. Another way of expressing the condition is that d, the coefficient of n^4, should be $= 0$, the actual error being $\frac{3}{4}d$. As a matter of fact, in most practical cases d would be very small and the final mean but little in error if the assumptions upon which the final mean method is based were correct.

These underlying assumptions are two, namely, that the tidal current varies according to a fair curve and that all runs back and forth are made at the same speed.

Every one who has often plotted results of speed trials in a tideway will have encountered results which could be explained only on the theory that the tidal current varied by fits and starts rather than according to a fair curve.

It is sometimes assumed that the tidal current varies from maximum to minimum in a manner such that a curve of tidal strength plotted on a base of time would be a curve of sines. This is perhaps a reasonable approximation to the general outline of the curve, but observations of actual strengths of tidal currents appear to show that they vary erratically and would seldom plot as a fair curve closely approximating a mathematical curve of sines.

A more serious error than that due to tidal current is liable to result from the fact that successive runs of a group are not made at the same speed. This is a matter of practical experience. It is very unusual, indeed, for four successive runs to be made over a measured course where the revolutions per minute, if accurately determined, do not vary appreciably. If the speed were constant, the revolutions should not change. Suppose now four successive runs were made aiming at a uniform speed of ten knots, while the actual speeds were 9.72, 10.24, 10.16, 9.88. The true average speed would be ten knots, but the final mean of the four speeds above would be 10.1 knots. This is quite a large error. In the above I have not taken account of the tide. The error is not affected if the tide is such that the final mean would eliminate the tidal error if the runs were made at constant speed. For instance, suppose the tidal currents were in knots .61, .74, .89, 1.06. For ten knots true speed the apparent speeds would be 10.61, 9.26, 10.89, 8.94. The final mean of these four speeds is 10 knots, as it should be. But if the true speeds of the successive runs were as given above, the apparent speeds after making allowance for currents, would be 10.33, 9.50, 11.05, 8.82. The final mean of these is 10.1 knots, as before.

Evidently, then, as the final mean method is equivalent to giving the two middle runs of a set of four a weight of three as compared with a weight of one for the first and last runs, when it is used for speed it should in theory be also used for revolutions and power. Thus, if a middle run of a series of four is made at a true speed above the average the excess speed in determining the average speed is given a weight of 3. This run will show excess power and revolutions, and if the average power and revolutions are properly to correspond with the average speed by the final mean method the power and revolutions should be given the same weight as the speed in determining the average. Practice in this respect appears to be somewhat variable. We often, but not always, find the final mean method used for revolutions. It appears to be seldom used for power.

4. Methods of Conducting Progressive Trials. — We seem warranted in concluding that when we attempt to get a spot on a speed and power curve by applying the final mean method to the data observed during a series of four runs, we by no means eliminate the probabilities of error. The question arises whether there are not better methods, or simpler methods equally good. We wish to determine curves — as accurate as possible — expressing the simultaneous values of speed, revolutions and horse-power. Now in any particular case we can usually determine the revolutions with great accuracy. We can determine the indicated horse-power with reasonable approximation, and with good indicators the error is as likely to be in excess as in defect. For twin-screw vessels, when the two engines show different revolutions during a run, the best we can do is to take the total indicated horse-power as corresponding to the average revolutions of the two engines. For any run we can determine the speed over the ground with ample accuracy, but owing to tidal current we cannot determine accurately the speed through the water. Now in plotting our results shall we plot power and speed on revolutions, or power and revolutions on speed, or perhaps speed and revolutions on power? A little consideration will show that there are real advantages in using revolutions as the independent variable, so to speak, and from the trial data plotting on revolutions separate curves of power and speed. For the revolutions of a run can and should be determined exactly to all intents and purposes.

Then by plotting our approximate data upon the correct revolutions we get rid of one element of uncertainty. We do not, for instance, plot a spot for power where the error is in excess over a spot for speed where the error is in defect. We will ultimately arrive at a more reliable relation between speed and power by determining first the most reliable relation between each and the accurately determined revolutions. Starting, then, with the basic idea that we will in the first place plot speed and power as ordinates upon revolutions as abscissæ, how should the progressive trial be conducted in order to determine most reliably the relation between power and revolutions?

We know from the Theory of Probabilities that if we wish to determine a single quantity — as, for instance, the value of a fixed angle — the best plan is to take as many observations as possible and use the average as the best obtainable approximation to the true value. Similarly, if we wish to determine a curve from experiment the best plan is to ascertain as many approximate spots as possible, plot them and draw the final curve as the average curve through the spots. Then to establish a curve of power on revolutions we should make numerous simultaneous determinations of power and revolutions, plot the results and draw an average curve through. To locate the curve of power as accurately as possible from a given number of runs, it would be better to have each run made at different revolutions. This would enable us to cover the curve closely with experimental spots. Here we encounter another weak point of the four-run final mean method.

Sixteen runs are necessary to determine four spots on a power curve, and four spots are insufficient for the accurate determination of a curve of power covering a wide range of speed. On the other hand, sixteen spots distributed at approximately equal intervals over the whole length of the curve will locate it with great accuracy. Each spot may be in error, owing to limitations on accuracy of any determination of indicated horse-power, but if the errors are as likely to be positive as negative a fair average curve through sixteen spots will practically eliminate the indicator errors. If the indicators have a constant positive or negative error no number of experimental spots will eliminate it. I conclude, then, that as to the relation between power and revolutions about sixteen simultaneous determinations of revolutions and indicated horse-power, made at approximately equal intervals of revolutions, will enable a satisfactory power revolution curve to be drawn. These observations need not necessarily be taken on the measured course, when the speed revolutions observations are being made. It is usual, however, to take the indicator cards while on the course. When the observing staff is adequate it is more convenient to make one job of it, and if the water on the measured course is somewhat shallow, so as to affect the results, it is desirable to determine everything under the same conditions. By doing this, too, we avoid the chance of the initial friction of the engines altering between two sets of runs, one to determine the power revolution relation, and the other the speed revolution relation. Finally, with an ample observing staff the time of a run over the measured course is generally of a length convenient for taking several sets of cards. There is, however, something to be said in favor of making runs off the course for determining power revolutions spots. With a small observing staff indicator cards can be taken more at leisure and given revolutions can be maintained until a sufficient number of satisfactory cards are taken, even if indicator accidents crop up. Again, as soon as good cards for a given number of revolutions are obtained the revolutions can be changed at once — up or down. This will not save much time at high speeds, but will at low speeds, so that the total time the staff must be kept at the indicators will be a good deal shorter. The preferable method really seems to depend in the end upon the observing staff available. With an ample staff of skilled observers, so that in addition to time and revolutions on the course three good cards can (barring accident) be obtained during each run from each end of each cylinder, it would seem advisable to make all observations on the measured course. With a small staff of observers, however, including many without good experience in such work, it would often be advisable to run separate trials, making the progressive power revolution trial before or after the speed revolution trial on the measured course.

Fig. 258 shows trial spots and final curve of power on revolutions as drawn from the trial of an armored cruiser.

Let us consider now the most suitable practical method of determining the speed-revolution relation from trials on the course. In the first place, no method will give reliable results unless we have a sufficient number of runs. Each experimental spot is necessarily and unavoidably somewhat in error. Hence, in order to get a reliable curve we must have so many spots and have them so close together that the accidental and erratic errors are practically eliminated by drawing a mean fair curve. There are two methods which may be used with confidence. The first is probably the most accurate and reliable, provided the trial is conducted with special skill along the lines described below. It is also adapted to the determination of the power revolution relation by the method just given. The second method is probably preferable for the usual run of trials.

Under the first method make a series of runs back and forth alternately with and against the tide and increasing or decreasing the revolutions by equal increments after each run. The **curve of** true speed then will fall midway between the two curves of apparent speed, one with and one against the tide. The advantages of this method are that if the curve of tidal variation is a fair curve and the trial skillfully run so that the interval between successive runs varies according to a fair curve, all spots of apparent speed will fall upon fair curves. Should, however, a spot be erratic, it will naturally fall off the curve and be given little weight in drawing the final curve of apparent speed. It is a very real advantage in such work to have a method of reducing the data such that bad spots show for themselves and are not incorporated in the final results. It is evident, however, that to get reliable curves of apparent speed we should have a sufficient number of spots for each curve. Not less than sixteen runs in all should be made. Figure 259 shows curves of apparent speed with and against the tide and the mean curve from the trial of an armored cruiser. All experimental spots are indicated.

There are some objections to the above method. One is, that at top speed, the most important part of the curve, we would have only one run, and the high speed part of the curve would not be defined so well as the lower portion. This difficulty should be overcome by making three runs at top speed — two in one direction, and one in the other, — and determining the final speed of the three by giving the middle run double the weight of the others. This is equivalent to taking the second mean of the three runs. The other objection to this method is that for thoroughly satisfactory results a trial once begun should be completely carried through without stopping. This sometimes introduces practical difficulties. A run may be lost through breakdown of the observing apparatus or interference of some other vessel while on the course. This is not a very serious objection, because it is found in practice that even if the intervals between the runs are somewhat erratic the curves of apparent speed can be drawn tolerably well. Another objection of the same nature is that if a trial is interrupted after five or six runs the results of these runs are of little value, as they are not sufficiently numerous accurately to define the part of the curve to which they refer, and a whole new trial has to be made.

The second recommended method of running a progressive trial is to make runs in groups of three,— 18 in all for a fast vessel, 15 for a vessel of moderate speed and 12 for a slow vessel. Each group should be made at a constant number of revolutions, as nearly as possible, the revolutions for the various groups covering the range desired. Then, taking for each group of three runs the second mean of speed and revolutions we have for a fast vessel six spots through which to plot a curve of speed and revolutions. This method in practice gives from each group of three runs a spot substantially as reliable as if four runs had been made. While it has the advantage, as compared with the four-run method, of giving more spots on the curve for a given total number of runs, it also has the advantage of beginning consecutive groups with runs in opposite directions. That is to say, if one group began with a run to the north the next group will begin with a run to the south. This is a desirable condition, as tending to eliminate some of the errors due to tidal current. This method has the advantage of requiring less skill and care in the conduct of trials, and each group of three runs stands by itself and is not wasted in case it is necessary to stop the trial. It is not quite so accurate as the method previously described, but the difference in accuracy would not be appreciable in the majority of cases. A practical advantage is that it does not require readjustment of throttles and links after each run in order to change the revolutions. This adjustment, in order rapidly to change revolutions by a definite amount, is by no means the simple matter it might appear at first thought and requires quick and accurate work in the engine room.

If there were no variations of tidal current between runs both methods above described would be theoretically exact. It is evidently desirable to time the progressive trial so that during it there shall be as little variation of tidal current between runs as possible. Now, when the tidal current is at a maximum, whether ebb or flow, the variation of current is at a minimum, while about the turn of the tide the rate of variation is about at a maximum. This statement would be exactly true if the curve of tidal current plotted on time was a curve of sines, as often assumed, and is substantially correct even as applied to actual tidal currents, varying by leaps and bounds rather than with definite progression. Then a progressive trial should always be run during the strength of one tide. A trial can generally be run in four hours or less, and so should, if practicable, be begun about an hour and a half after the turn of the tide. Circumstances often render this inconvenient or impossible, and weather conditions frequently cause the turn of the tide to come before or after the time fixed by tide tables, but the best time for a trial should be used unless there are good reasons to the contrary.

So far as accuracy of results is concerned it makes no difference whether we begin with the low speeds and work up or with the high speeds and work down. It seems advisable, however, as a rule to begin with the top speeds and work down. With clean fires and fresh men the top speed can be obtained and maintained with more ease than after several hours of running. Also, if the trial is spoilt by a breakdown it is more apt to come during the high speed runs, and if a breakdown must come it is better to have it come early than late.

There may be mentioned here some minor points in connection with the conduct of trials which tend to produce accurate and satisfactory results. It is desirable after a run to shift revolutions promptly to the revolutions for the next run, if they are to be different. If there is a pressure gauge giving the pressure in the H.P. chest (beyond the throttle) it is easy by preliminary runs to establish a curve (or curves, if more than one valve gear setting is to be used) giving the relation between H.P. valve chest pressure and revolutions. Then it is necessary only to establish the proper pressure to insure that the revolutions are sufficiently near what is desired. Such a pressure gauge as above is apt to fluctuate violently unless its cock is nearly closed. Systematic handling of the ship when off the course is desirable. Each time when coming on the measured course the ship should have made a long straight run with the minumum operation of helm. For most trials about a mile is a convenient and desirable length for the straight run, and it much facilitates trials if in addition to buoys at the ends of the measured course, moored closely on the ranges, there are planted buoys in the line of the course a mile from each end. Suppose we have the course thus buoyed as indicated in Fig. 257. Before beginning the trial proper — while warming up — steam over the course as indicated in Fig. 257 by $ABCDEFCBGHA$.

When abreast the buoy D put the helm over to a moderate and definite angle, say 10 degrees. Steady the ship on the course EF which will cross the line of the course a little beyond the buoy C. While on this course note carefully the compass reading and determine the reading of the steering compass which will give the opposite course FE. Then when coming off the course at C after a run, put the helm over at once and steady the ship on the course FE. If the revolutions are to be changed for the next run the engine room force should immediately set to work on this. With skillful handling the new desired revolutions should be attained before the vessel is at E. If this is so, on reaching E abreast the buoy D put the helm over to 10 degrees. The vessel will, by the time she swings to the correct heading for the next run, be practically on the line of the course, requiring very little use of the helm to come dead on. If the revolutions are not adjusted by the time the vessel reaches E, she should at this point be steadied on the course EK, shown dotted in Fig. 257, and kept on this course until the revolutions are satisfactorily adjusted or the vessel has run so far that there will be ample time after turning finally to adjust the revolutions before the vessel reaches D. The methods are of course just the same at each end of the course.

To conduct a trial in this way requires quick communication and complete understanding between the deck and the engine room, but results will be distinctly superior to those obtained by more haphazard methods.

5. **Trial Conditions.** — It is customary to make progressive trials with clean bottoms under good conditions of wind and sea. For men-of-war the trial is generally made at normal load displacement. For merchant vessels the displacement is sometimes the average displacement to be expected in service, but generally a less displacement and at times a very light displacement.

The usual practice is at times criticised. As to men-of-war, for instance, it is alleged that they will never show in service such good results as upon trial. It is true that there is ever present the temptation to run trials at too light a displacement. This is largely due to the natural desire of those concerned to make the best showing possible. But the loss of speed in service due to increased displacment is apt to be exaggerated, particularly for large ships. More potent causes are rough water at sea, dirty bottoms, poor coal, or inability of the engineering personnel to get good power results. It is evidently desirable to have trials always run under uniform or standard conditions. The most easily attained standard trial conditions are obviously fair weather, smooth water and a clean bottom. From reliable results under such conditions the results which should be attained in service can be estimated with sufficient approximation until they can be ascertained by experience. As a general thing, however, progressive trials cannot, and are not expected to, show exactly what a ship will do in service. This requires service experience. They furnish data to enable the performance of the ship under standard conditions to be determined and compared with other vessels, and in case the performance is poor careful progressive trials will not only determine that fact, but as a rule, upon analysis, indicate the line that should be followed to obtain improvement.

36. Analysis of Trial Results

1. **Components of Indicated Horse Power.** — Figure 260 shows a curve of speed and power for the U. S. S. *Yorktown*, the powers as ordinates being plotted over the speeds as abscissæ.

The power is the indicated horse-power developed in the cylinders of the engines. We know that only a fraction of this power is finally utilized to propel the ship and it is important to gain some idea of the distribution of the remainder.

The engine itself absorbs a certain amount of power through its own friction. This friction is usually classed under two heads, namely, " initial " or " dead " friction, due to tightness of pistons, valves, glands, bearings, etc., and " load " friction, or the friction due to the load upon the bearings and thrust block.

The power required to work feed, air, circulating and bilge pumps, driven from the main engines, is usually classed with the initial friction. For reciprocating engines, the power P delivered to the propeller is the original indicated horse-power less the power as above absorbed by friction. For turbine engines the power is usually determined from the twist of the shaft, measurements being taken astern of the thrust block. All of this shaft horse-power is delivered to the screw except what is wasted in friction of line bearings, stern tubes, and outward bearings, if any. This is usually so small that the shaft horse-power is assumed to be the same as the propeller power P.

Of the propeller power P a portion is wasted in friction and slip of the propeller. The remainder is used in developing thrust horse-power. Also there is added here a certain amount of power derived from the wake which also appears as thrust horse-power. Of the thrust horse-power a certain amount is used to overcome the augmentation of resistance of the ship due to the suction of the propeller, and the remainder is the effective horse-power, the net power required to drive the ship.

The above components of the I.H.P. vary widely. The initial friction will absorb from as low as 3 or 4 per cent of the power in large well-adjusted engines with independent air and circulating pumps to 10 per cent or more in the case of machinery badly adjusted with air and circulating pumps driven off the main engines.

The load friction is usually taken as about 7 per cent of the remainder obtained by deducting the inital friction power from the original I.H.P. With well-lubricated engines it is generally somewhat less. Investigations of the shaft horse-power of reciprocating engines by means of torsion meters have shown as much as 92 per cent of the indicated horse-power delivered to the shaft, involving a loss of but 8 per cent for both initial and load friction. Engines seldom run any length of time with excessive load friction. It promptly causes hot bearings.

The ultimate distribution of the propeller power — the shaft horse-power for turbine jobs — is a question of the efficiency of the propeller, the wake factor and the thrust deduction.

It is evident from what has gone before that as a reasonable working approximation we may assume that for a reciprocating engine of high-class workmanship about 90 per cent of the indicated horse-power is delivered to the propeller when independent air and circulating pumps are fitted, and about 85 per cent of the indicated power when all pumps are driven off the main engine.

Accurate trial results can be analyzed to give an approximation to the resistance of the ship, and hence efficiency of propulsion, etc., but these quantities can be estimated directly with sufficient accuracy and with much less labor by methods already given. It is very desirable, however, to determine accurately the initial friction of an engine, as then we know with close approximation the propeller power, P, and this power is an essential factor of the propeller design. Hence we will now consider in detail the initial friction of an engine and methods for determining it from progressive trial results.

2. Initial Friction Determined by Curves Extended to Origin. — Mr. William Froude, the pioneer investigator of this question, defines initial friction as " the friction due to the dead weight of the working parts, piston packings, and the like, which constitute the initial or low speed friction of the engine." The initial friction, or internal resistance, is generally regarded as constant throughout the range of speed and power of the engine, thus differing from the load friction, which is generally regarded as absorbing a uniform fraction of the power developed. As a matter of fact, it seems altogether probable that the internal resistance varies slightly with power and revolutions, but the variation is probably so small as long as bearings run cool that we are justified in ignoring it.

There is no doubt that the internal friction will alter materially from time to time, due to changes in tightness of various parts. The problem under consideration, however, is the determination of the initial friction at a given time. If the frictional resistance is constant the power absorbed by it will be proportional to the revolutions, so that if we denote by I_f the horse-power absorbed by initial friction and R denotes revolutions, we have $I_f = R \times$ (a coefficient), where the coefficient at a given time for a given engine is constant. Suppose we denote the coefficient by C_f, then $I_f = C_f R$. Now analysis and consideration of the various absorbents or components of the indicated horse-power, such as the power utilized to propel the ship, the power wasted by the propeller, the power absorbed in load friction, etc., show that they all, except I_f, must vary as some power of the revolutions greater than unity. This being the case, it follows that if I denote the indicated horse-power at revolutions R, we may write $I = C_f R + \phi(R)$, where we know that $\phi(R)$ is some function of the revolutions which varies always as a power of R greater than unity. If, then, we plot a curve of I on revolutions, as we approach the origin the curve of I will approach the straight line $I_f = C_f R$, and at the origin will be tangent to this line. Hence C_f can be determined from the inclination at the origin of the curve of I plotted on R.

Figure 261 shows for the U. S. S. *Yorktown* a curve of indicated horse-power plotted on revolutions, the curve being extended to the origin and the tangent at the origin being drawn in. It is desirable in plotting this curve to draw, as shown, a similar symmetrical curve in the third quadrant joining the real curve in the first quadrant to the imaginary curve in the third quadrant at the origin where there is a point of inflection. This facilitates drawing a curve which has the proper direction at the origin. Then drawing the tangent at the origin we determine the line for $I_f = C_f R$, and taking at any point the simultaneous values of I_f and R we have $C_f = \dfrac{I_f}{R}$.

Another method is to plot a curve of I divided by R in the first quadrant and a symmetrical curve in the second quadrant. Such a curve will not pass through the origin but cut the axis of $R =$ zero at a point above the origin. Its ordinate here is evidently C_f. The ordinates of the curve of $\dfrac{I}{R}$ bear a constant ratio to the ordinates of the curve of mean effective pressure.

It is customary to reduce the initial friction or internal resistance of an engine to equivalent mean effective pressure in the low pressure cylinder or cylinders. This is the most convenient and probably the most reliable way of comparing engines of different types and sizes as regards internal resistance.

Let n denote the number of L.P. cylinders, d the diameter of each in inches, s the stroke in inches, p_m the mean effective pressure in pounds per square inch reduced to the L.P. cylinder area and R the revolutions per minute. Then

$$I = \frac{n \frac{\pi}{4} d^2 p_m \times \frac{2s}{12} \times R}{33000} = \frac{n d^2 s R p_m}{252100}.$$

At the limit $I = I_f = C_f R$. If p_f denote mean effective pressure equivalent to internal resistance reduced to L.P. area, at the limit

$$p_m = p_f \text{ or } C_f = \frac{nd^2 s p_f}{252100} \text{ or } p_f = \frac{252100\, C_f}{nsd^2}.$$

It is seen from the above that when we have once determined a reliable value of C_f we can readily obtain the corresponding value of the mean effective pressure in the low pressure cylinder from the known data of the engine. If we could determine with accuracy the curve of indicated horse-power for a given engine to a very low number of revolutions the above method of determining internal resistance would leave little to be desired. However, we meet here with a number of practical difficulties. If we determine simultaneous values of speed, power and revolutions, which is the usual practice in progressive trials, it is found that the low speed trials over a measured mile are very tedious. If we avoid this trouble by determining in free route at the lowest speeds the horse-power and revolutions only, we still encounter difficulties. No reciprocating engine will run at all below a certain speed, and as it approaches the limiting speed at which it will stick, its action becomes somewhat erratic and uncertain. It is true that the less the friction the lower the revolutions at which the engine will stick, and that this is a rough measure of the initial friction; but even the smoothest running engines will seldom run steadily down to a speed sufficiently low to enable the internal resistance to be determined with accuracy by a curve extended to the origin. For determining very low speed powers of engines which use high pressure it is necessary to use special weak indicator springs, otherwise the indicator diagrams have such a very small area that the determination of the power is very uncertain. If, instead of determining a curve of power and revolutions for the ship under way we determine the same thing for the vessel tied up at the dock, we will get larger indicator cards and the engine will turn over at a slightly lower number of revolutions, but even then the results generally leave something to be desired.

Torsion meter apparatus has been designed of late years to measure the power being transmitted by a shaft by determining the twist of the shaft. If we measure shaft horse-power by a torsion meter and simultaneously indicate the engine, we can determine the total frictional resistance of the engine, the power absorbed by friction being of course the difference between the indicated horse-power and the shaft horse-power. With accurate data this would probably be the most nearly exact method of determining the initial friction of the engine and would have the incidental advantage of enabling the load friction to be determined as well, but the accuracy of torsion meters at low speeds and powers is not sufficient to enable this method to be made use of except perhaps in very exceptional cases. It is evident that we need some method of obtaining the desired result from an ordinary curve of power and revolutions which does not go below a speed and power for which the data may be readily obtained and regarded as fairly reliable. It is natural to ask whether there is any inherent feature or property of curves of horse-power which would facilitate the determination of the internal friction. Mr. William Froude worked on these lines. He plotted a curve of indicated thrust upon the speed of the ship in knots, carrying the curve down as low as possible. Indicated thrust is a thrust which at the speed of the propeller will absorb the indicated horse-power. At zero speed and zero revolutions the curve of indicated thrust, whose ordinates are proportional to $\frac{I}{R}$, will cut the axis of thrust at a distance above the origin proportional to the initial friction. To pass from the lowest point of his curve of indicated thrust, determined by observation, Mr. Froude made use of an essential property of these curves. He assumed that at these low speeds the resistance of the ship varied as the 1.87 power of the speed, and that all other losses except the initial friction loss were constant fractions of the power absorbed by resistance. It would follow that the curve of indicated thrust in the vicinity of the origin is a parabola of the 1.87 degree whose ordinate at zero speed is proportional to the initial friction.

Now, referring to Fig. 262, if the curve therein indicated is a parabola of the 1.87 degree it follows that the tangent at the point P will cut the horizontal tangent through the lowest point A at a point M, so that AM divided by AN is equal to $\frac{.87}{1.87}$. Mr. Froude, then, having drawn his curve of indicated thrust to as low a speed as he could from the data, next drew the tangent at its extremity as KB in Fig. 263, and dividing OL at H so that $\dfrac{OH}{OL}$ equals $\dfrac{.87}{1.87}$ he set up HB to intersect the tangent at K in the point B. A horizontal line, then, through B cuts the axis of thrust at the point T, and OT is the indicated thrust corresponding to the initial friction. This method makes use of a property of the curve, but as a matter of fact, it is hardly so reliable in practice as the method of extending the curve of indicated horse-power to the origin and setting off the tangent to it. While the low speed resistance of the ship would be reasonably close to the 1.87 power of the speed this is still an approximation, but the principal objection to this method is that it requires a tangent to be drawn at the low speed extremity of the curve of indicated thrust. The difficulty of obtaining reliable values for this curve at the lowest speed have been pointed out and it follows, apart from the difficulty of drawing an accurate tangent at the extremity of any curve, that an error in the low speed spot would throw out the low speed tangent and introduce material errors.

3. Initial Friction Deduced from Low Speed Portion of Power Curves. — The question arises, then, whether we cannot make use of some inherent property of curves of horse-power which will enable us to determine the initial friction with reasonable accuracy without it being necessary to carry any curve to the origin. We know that the frictional resistance of a ship varies about as the 1.83 power of the speed, so that the horse-power absorbed by frictional resistance varies as the 2.83 power of the speed. The power absorbed by wave making varies as a higher power than the cube of the speed. The practical result is that at low speeds, when there is almost no wave resistance, the total effective horse-power will vary as a somewhat lower power of the speed than the cube, whereas at high speeds it will vary as a higher power of the speed than the cube. There is then some point at moderate speed where the effective horse-power is varying as the cube of the speed.

Consider now the propeller. For a given slip the power absorbed by a propeller varies as the cube of the revolutions, or for constant slip as the cube of the speed. It follows, then, that starting from a very low speed, where the effective horse-power is varying at a lower power than the cube, the slip of the propeller falls off until we reach the speed at which the effective horse-power varies as the cube of the speed. At this point the slip of the propeller reaches a minimum beyond which it increases. The efficiency of the propeller at the point where the slip reaches a minimum will be constant, and the power delivered will vary as the cube of the speed or as the cube of the revolutions. Also all losses will vary as the power delivered to the propeller or as the cube of the revolutions, except the initial friction loss. Hence, at the point of minimum slip where the slip remains constant for a minute interval the following formula will express exactly the indicated horse-power:

$$I = C_f R + c R^3.$$

For some little distance on either side of the point of minimum slip the above formula will give a reasonably close approximation to the facts, especially for the speeds below the point of minimum slip. Now C_f and c in the above equation are both unknown, but from the curve of indicated horse-power plotted on revolutions we can determine any number of simultaneous values of I and R, and for each pair of such values we can draw a straight line on axes of c and C_f, constituting a focal diagram. If the equation above applies throughout to the curve of indicated horse-power and C_f and c were constant, it would follow that this diagram would have a perfect focus. Now we know that the above equation does not apply to the upper part of the curve of horse-power at which the indicated horse-power generally varies as a very much higher power than the cube. It seems reasonable from the nature of the case, however, that this equation should be fairly approximate over a tolerably wide range of the lower speeds, and that if we draw for this range a series of lines C_f and c, they should all pass reasonably close to a common point; in other words, should form a reliable focal diagram. Investigation of practical cases shows that we do have such a focal diagram. The methods of calculation are very simple. The table below shows the calculations for the *Yorktown*, and Fig. 264 shows the diagram for the *Yorktown*.

CALCULATIONS TO DETERMINE INITIAL FRICTION COEFFICIENT OF U.S.S. YORKTOWN.

V	5	6	7	8	9	10	11	12
I	96	143	205	286	393	536	728	986
R	44.8	53.7	62.6	71.5	80.6	90.0	99.7	110.0
R^2	2007	2884	3919	5112	6496	8100	9940	12100
If $c=0$, $C_f = \dfrac{I}{R}$	2.143	2.663	3.275	4.000	4.877	5.955	7.302	8.964
If $C_f=0$, $c = \dfrac{I}{R^3}$	.001068	.000923	.000836	.000782	.000751	.000735	.000735	.000741

It is seen that taking the four lines corresponding to speeds of 5, 6, 7 and 8 knots we get an excellent focal diagram. The 9-knot line does not give quite such a good intersection and the 10-knot line leaves the focus entirely. This is typical of such diagrams, and the lines themselves show very clearly which should be used and which should not be used in determining the focal point. C_f from Fig. 264 for the *Yorktown* is equal to 0.96, while its value from Fig. 261, taken from the curves extended to the origin, was made .95 a number of years ago, soon after the trial of the *Yorktown* in 1889. The practical agreement of the results of the two methods might seem in favor of the method of extension to the origin, which is the simpler. But special care was taken on the *Yorktown* trial to obtain reliable power data at very low speed, and the results obtained by extending her power curve to the origin can be regarded with more confidence than those of subsequent trials of other vessels where it was found practically impossible to extend the power curves to the origin with certainty. It was this fact which impelled a search for a better method. Figures 265 to 269 show initial friction diagrams for the United States ships *Alabama*, *Kearsarge*, *Massachusetts* and *Maine* and the revenue cutter *Manning*. The mean effective pressure in the L.P. cylinders equivalent to initial friction and the percentage of maximum power absorbed by initial friction are given below for these vessels and the *Yorktown*.

Name of Ship.	Yorktown.	Alabama.	Kearsarge.	Massachusetts.	Maine.	Manning.
Mean effective pressure in L.P. equivalent to initial friction..............	1.61	2.07	1.63	1.77	3.20	2.52
Per cent of max. power absorbed by initial friction.................	4.28	4.94	3.67	4.33	7.44	6.59

The above vessels were all given careful trials and the results are as reliable as will usually be obtained. While the diagrams show lines for successive speeds, successive values of revolutions could have been used as well, and in fact the method can be readily applied to a curve of power and revolutions where the speed is not known. It is seen that in every case there is an excellent focus formed by the lines for the lower speeds, except in the case of the *Maine*, where the focus is not so well defined as would be desirable. The generally satisfactory determination of the focus in accordance with theoretical reasoning may be regarded as fairly strong evidence in favor of the method outlined above. To produce direct evidence for this method we can apply it to a case where the internal resistance is accurately known by some other method. The *Yorktown* was one such case. Fortunately, however, we can produce stronger cases. In the transactions of the Society of Naval Architects and Marine Engineers we find two cases of determinations of speed and power of double-ended ferry boats with a propeller at each end. Three curves are given for each case, one curve for both screws in use, one for only the stern screw in use, the bow screw being removed, and one for only the bow screw in use. One case was that of the *Cincinnati*, the data for which can be found in a paper by F. L. DuBosque, in the volume for 1896, and the other case was that of the *Edgewater*, the data for which can be found in a paper by E. A. Stevens in the volume for 1902. Fig. 270 reproduces the curves of power plotted on revolutions for the *Cincinnati* and Fig. 271 the similar curves for the *Edgewater*. It is seen that the three curves for each boat differ radically from each other, owing to differences of propeller efficiencies, etc., but it is evident that for each vessel the internal friction of the engine should not vary much for the three conditions, since the engines, shafting, etc., were the same and the only factors affecting frictional resistance were the presence or absence of one screw and the variations of initial friction between trials. Figures 272 and 273 show the frictional focal diagrams for the *Cincinnati* and *Edgewater* as deduced from Figs. 270 and 271 and the curves of speed and revolutions. The original observations for the *Cincinnati* do not extend to quite so low a speed as desirable for the initial friction determination, but it is seen that the several cases, in spite of the radical differences in the curves of power, give fairly satisfactory foci in adequate agreement. The average value of C_f for the *Edgewater* is .738, the highest value being 5.0 per cent above and the lowest value 7.2 per cent below the average. Similarly for the *Cincinnati* the average value of C_f is .897, the highest value being 8.7 per cent above and the lowest value 13.0 per cent below the average.

I think, then, it may be safely concluded that the Focal Diagram method outlined above will give a definite determination of the initial friction which, with good data, may be expected to be within 10 per cent of the truth. This approximation is ample for practical purposes, since at the higher speeds the whole initial friction power is but a small percentage of the total. It will be observed that the focal points are simply spotted by eye on the focal diagrams. The theoretical most correct focus of such a diagram can be determined by Least Square methods at the expense of not very much time and trouble. Since, however, the results obtained are approximate in any case, we gain no real additional accuracy by the extra calculations.

4. Determination of Efficiency of Propulsion from Trial Results. —The efficiency of propulsion being the ratio between effective and indicated or shaft horse-power we need to know the effective horse-power in order to determine it for any speed.

The effective horse-power may be that of the bare hull or include the appendages. In either case, given the curves of E.H.P. and of I.H.P., the determination of a curve of efficiency of propulsion is simple and obvious.

Since initial friction absorbs a greater proportion of the power at low speeds we may expect to find for vessels with reciprocating engines the efficiency of propulsion falling off rapidly at low speeds. If propeller efficiency, wake factor, etc., were constant, the maximum efficiency of propulsion would always be found at top speed, but propeller efficiency varies with slip, which is not constant as speed changes, and the wake fraction also varies with speed. Hence, we frequently find the maximum efficiency of propulsion below the maximum speed. But in most practical cases unless cavitation sets in the efficiency of propulsion does not change much either way for several knots below the maximum speed.

If curves of E.H.P. are deduced from experiments with a model of the ship the resulting efficiencies of propulsion are of course more reliable than those obtained from estimated curves of E.H.P. If, however, model experiments are not available for a vessel for which we have reliable power data we should always estimate curves of E.H.P. from the Standard Series diagrams (Figs. 81 to 120) and deduce curves of what may be called nominal efficiencies of propulsion. Such nominal efficiencies for vessels of a definite type are, when dealing with a new vessel of the same type, almost as useful as if they were derived from model tests.

5. **Analysis for Wake Fraction and Thrust Deduction.** — When considering the question of wake in Section 28 we saw how from the propeller power P and the revolutions and speed we could estimate the wake fractions by a curve of S from experiments with a model of the propeller or by the standard curves of S (Figs. 230 to 233).

As the values of P and S used are at best experimental and approximate, the most that can be hoped for wake fractions thus determined is that they will be reasonably good approximations.

If there is cavitation the method fails, and there is reason to believe that propellers with blunt or rounding leading edges cavitate without it being discovered. The effect of slight cavitation, or in fact of any failure of the Law of Comparison, is to cause the wake deduced from Figs. 230 to 233 or by similar methods to be less than the real wake. This possibility should always be borne in mind when analyzing trial results for the determination of wake. Theoretically we can determine thrust deduction factors from analysis of trial results in connection with accurate model results for ship and propeller tested separately.

For
$$1 - t = \frac{\text{E.H.P.} (1 - w)}{eP},$$

where E.H.P. is effective horse-power of ship, e is propeller efficiency, P is propeller power and w is wake fraction. In practice, however, since every quantity on the right of the above equation is estimated or only approximated, the thrust deduction factors thus determined are seldom reliable.

CHAPTER V

THE POWERING OF SHIPS

37. Powering Methods Based upon Surface

1. Rankine's Augmented Surface Method. — The methods that have been proposed and used to estimate the power required to drive a given ship at a given speed are many and various. One of the earliest English methods which broke away from the rule of thumb and attacked the problem in a logical and scientific way was Rankine's Augmented Surface method, brought out some fifty years ago. Rankine assumed that in a well-formed ship the resistance was wholly frictional, the water flowing past the ship with perfect stream motion and the frictional resistance varying as the square of the speed.

But with perfect stream motion the average relative velocity of flow over the surface would be somewhat greater than the speed of the vessel with reference to undisturbed water, and Rankine developed elaborate mathematical methods for determining an "Augmented" surface such that its frictional resistance at the speed of the vessel, neglecting stream motion, would be the same as the actual frictional resistance of the real surface of the ship when there was perfect stream motion. Rankine assumed .01 as a coefficient of friction, so by his method we would have Resistance = .01 V^2 × Augmented Surface. We know now that Rankine's fundamental assumptions were wrong and would involve results vastly more erroneous in practice than the use of the actual surface instead of the slightly greater augmented surface. In his time, however, there were few fast ships, and the assumption that resistance was wholly frictional was not so much in error as it would be now. Furthermore, little was known of the actual coefficients and laws of frictional resistance, as William Froude's epoch-making experiments on the subject were subsequent to 1870. So Rankine's neglect of all resistances but friction was to some extent made up by his overestimate of the friction. The calculation of the Augmented Surface was, however, not easy, and for many years Rankine's method has been obsolete.

2. Kirk's Method. — A method of estimating power was brought out by Dr. A. C. Kirk of Glasgow nearly thirty years ago, which though resembling closely Rankine's method in basic underlying principles, is much simpler and easy of practical application. Dr. Kirk devised in the first place a method of approximation to the wetted surface S. He then assumed that the resistance would vary directly as the square of the speed and the indicated horse-power as the cube of the speed, using the formula $I = \dfrac{kSV^3}{100000}$ where I is indicated horse-power, V is speed in knots, S is wetted surface in square feet and k is a coefficient which must be fixed by experience.

Kirk made $k = 5$ for merchant ships of ordinary proportions and efficiency, while for fine ships with smooth clean bottoms and high propulsive efficiencies it was as low as 4 and for short broad ships as high as 6.

For the low speed cargo vessel for which Kirk devised and recommended his method it has many excellent features.

For such vessels the residuary resistance is usually not a large proportion of the whole, and up to 11 or 12 knots the I.H.P. does vary approximately as the cube of the speed.

Then the coefficient k was fixed, not by preconceived ideas or reasoning as to what it ought to be, but by experience of what it had been on other similar ships. Hence, Kirk's method is sound in principle. The main objection to it is that it is of little value for fast vessels, and even for the 10 to 12 knot cargo boat the coefficient k is apt to vary erratically.

3. Coal Endurance Estimated from Surface. — The principle of Kirk's method may be utilized to advantage for estimating the low speed endurance of vessels of war. Such vessels, whatever their full speed, usually make passages at a moderate speed of 10 to 12 knots in order to save coal or gain endurance. At such speeds the I.H.P. varies approximately as the cube of the speed V and as the wetted surface which is proportional to $\sqrt{DL}$. Hence, I.H.P. varies as $V^3 \sqrt{DL}$

Now at these low speeds the coal burnt for all purposes per I.H.P. varies inversely as some power of the speed and may be assumed to vary approximately as $\frac{1}{V^2}$.

Hence, coal per hour varies as $\frac{V^3 \sqrt{DL}}{V^2}$ or as $V\sqrt{DL}$.

Hence, coal per mile varies as $\frac{V\sqrt{DL}}{V}$ or as $\sqrt{DL}$.

Hence, miles steamed per ton of coal vary as $\frac{1}{\sqrt{DL}}$.

So if m denote the miles steamed per ton of coal and K_0 a coal coefficient, we have $m = \frac{K_0}{\sqrt{DL}}$. If the approximate assumptions above were exact K_0 would be constant for all ships and speeds. In practice K_0 varies from ship to ship and with the speed of a given ship. It increases from a very low speed up to a maximum value — nearly always for a speed below 10 knots which is the most economical speed for the ship.

For speeds beyond the most economical speed K_0 falls off steadily. Fig. 274 shows curves of K_0 for some United States battleships, the average of the sister ships *Kearsarge* and *Kentucky*, the *Wisconsin* and the *Oregon*.

These curves are averaged from consumption at various displacements with all kinds of coal, under all conditions of bottom and of weather and hence are from average service results. The *Wisconsin* data was not complete enough to make a reliable final average. On a given passage a vessel may well show values of K_0 twenty per cent above or below the average, with the varying conditions as respects quality of coal, state and management of the machinery, foulness of the bottom and the weather.

On the voyage of the United States Atlantic fleet around the world the *Kearsarge* and the *Kentucky* showed an average K_0 for 10 knots of 6900 as against about 7130 in Fig. 274. The *Wisconsin*, however, which has a 10-knot K_0 of only 6910 in Fig. 274, showed an average value of 7600 in the voyage around the world, the values on the different legs varying from 7300 to 7900. For the whole fleet the average value of K_0 was about 7200 at 10 knots. This figure may be regarded as fairly typical of large battleships with reciprocating engines, though it will be found that it will give such vessels endurances under average service conditions far below those usually credited to them in naval handbooks.

A flotilla of United States destroyers on its way from the United States to the Philippines via the Suez Canal some years ago showed an average value of K_0 at 10 knots of 5000. Merchant vessels designed for only 10 knots naturally show much larger values of K_0. Thus a large 10-knot naval collier on a voyage from Hampton Roads to Manila showed an average K_0 of nearly 13,000. An 18,000-ton ten-knot freighter in the Atlantic trade showed about 12,000 in three passages under moderate weather conditions, while on a passage made in exceptionally heavy weather throughout, its K_0 dropped to less than 9000.

4. **Admiralty Coefficients.** — Perhaps the method most used in the past for powering ships has been the Admiralty Coefficient method. Here again the basic assumptions are that the resistance is all frictional and the I.H.P. varies as the cube of the speed. The wetted surface is not used directly, however. For similar ships the wetted surface varies as the square of the linear dimensions, or as $D^{\frac{2}{3}}$ where D is displacement in tons, or as M where M is area of midship section in square feet. Hence we write

$$I = \frac{D^{\frac{2}{3}} V^3}{C_1} \text{ or } I = \frac{M V^3}{C_2},$$

where I is indicated horse-power, V is speed in knots, C_1 is the "displacement" coefficient and C_2 is the "midship section" coefficient.

It is evident from the above that $\frac{C_1}{C_2} = \frac{D^{\frac{2}{3}}}{M}$, so that for a given ship $\frac{C_1}{C_2}$ is constant throughout the range of speed. But for dissimilar ships the ratio between C_1 and C_2 is different, so that two ships on trial may show the same values of the displacement coefficient and very different values of the midship section coefficient, and *vice versa*.

In England, the displacement coefficient has been regarded as the most reliable, that is, as showing less change with variation of type of vessel. In France, on the contrary, the reciprocal of the midship section coefficient is largely used. It is evident, however, that any formula based upon the assumption that resistance varies as the square of the speed must be unreliable for high speeds unless there is available a large accumulation of data from trials of fairly similar high speed vessels. In such case, in spite of the faulty assumption, it may be possible to select a suitable coefficient.

It is apparent, however, that the Admiralty coefficients ignore a number of factors which have great influence upon resistance. For instance, both coefficients ignore the length and the longitudinal coefficient, — factors which are sometimes of enormous importance.

So, in spite of the long use that has been made of the Admiralty Coefficient method, it must be regarded as reliable only when on the well-beaten track. Reliable trial results from a number of vessels of different types will give Admiralty coefficients which vary widely.

When it is necessary to fix upon the coefficient to adopt when powering a new vessel, much experience and good judgment will be needed.

38. The Extended Law of Comparison

1. **Deduction of Extended Law of Comparison.** — The most accurate method known at present for the estimation of the resistance of a full-sized ship is to determine the resistance of a model of it and by using the Law of Comparison deduce the resistance of the full-sized ship.

Evidently, then, we may regard a full-sized ship whose trial results we know as a model and power similar ships from its trial results. Thus, suppose we have a ship of displacement D whose resistance is R at speed V, whose effective horse-power is E, indicated horse-power I and efficiency of propulsion e.

For a similar ship at corresponding speed let us denote the quantities enumerated above by D_1, R_1, V_1, I_1 and e_1.

We know by the Law of Comparison that

$$\frac{R}{R_1} = \frac{D}{D_1}, \quad \frac{V}{V_1} = \sqrt{\frac{L}{L_1}} = \sqrt[6]{\left(\frac{D}{D_1}\right)^3} = \left(\frac{D}{D_1}\right)^{\frac{1}{6}}.$$

Whence
$$\frac{E}{E_1} = \frac{RV}{R_1V_1} = \left(\frac{D}{D_1}\right)^{\frac{7}{6}},$$

and if $e = e_1$, which should be the case with sufficient approximation,
$$\frac{E}{E_1} = \frac{eI}{e_1I_1} = \frac{I}{I_1} = \left(\frac{D}{D_1}\right)^{\frac{7}{6}},$$

This is the Extended Law of Comparison, so called. We may express it by the statement that for similar models at corresponding speeds $\frac{I}{D^{\frac{7}{6}}}$ is constant.

2. Application of Extended Law of Comparison. — There are various methods of plotting the trial data of a ship so that by using the Extended Law of Comparison it can be applied to new designs. A simple method is to plot a curve of $\frac{I}{D^{\frac{7}{6}}}$ over values of $\frac{V}{\sqrt{L}}$. This eliminates the size factor. Thus, Fig. 275 shows a curve of $\frac{I}{D^{\frac{7}{6}}}$ for the U. S. S. *Yorktown* plotted on values of $\frac{V}{\sqrt{L}}$.

The *Yorktown* is of 230 feet mean immersed length, of 1680 tons' trial displacement and made about 17 knots on trial. Suppose we wish, from Fig. 275, to determine the necessary I.H.P. for a vessel similar to the *Yorktown*, 289 feet long, of 3333 tons' displacement, and to make 17 knots. Then for the 289-foot vessel
$$\frac{V}{\sqrt{L}} = \frac{17}{\sqrt{289}} = 1.$$

From Fig. 275 when
$$\frac{V}{\sqrt{L}} = 1 \qquad \frac{I}{D^{\frac{7}{6}}} = .415.$$

Also
$$(3333)^{\frac{7}{6}} = 12{,}881,$$

whence for the 289-foot vessel to make 17 knots
$$I = .415 \times 12{,}881 = 5345.$$

This is very simple, but for practical work it is convenient to plot our data a little differently. The curve of $\frac{I}{D^{\frac{7}{6}}}$ in Fig. 275 is quite steep and varies a great deal as we pass from low to high speeds.

So let us use instead a curve of

$$\frac{I}{D^{\frac{2}{3}}} \div \left(\frac{V}{\sqrt{L}}\right)^3 = \frac{I}{V^3} \times \frac{L^{\frac{3}{2}}}{D^{\frac{2}{3}}} = N \text{ say.}$$

Then $I = N \dfrac{V^3 D^{\frac{2}{3}}}{L^{\frac{3}{2}}}$. This is a convenient form. We may call N the Extended Law of Comparison coefficient.

Figure 276 shows curves of N as deduced from trial results for the *Yorktown* and several other vessels. The curves are numbered, and the dimension and proportion of the corresponding vessels are given below:

No.	Name.	Type.	Length L, feet.	Displacement D, tons.	$\frac{L^{\frac{3}{2}}}{D^{\frac{2}{3}}}$	Longitudinal Coeff.	$\left(\frac{L}{100}\right)^3$	Beam l, feet.	Draught H, feet.
1	Yorktown	Gunboat	230	1680	.6023	.505	138.1	36.00	13.82
2	Manning	Revenue Cutter	188	1000.7	.8145	.605	157.7	32.83	12.33
3	Dahlgren	Torpedo Boat	147	138	5.680	.665	43.4	16.38	4.34
4	Commonwealth	Paddle Str.	437.0	5430	.4025	.626	64.7	55.00	13.04
5	Birmingham	Scout	420	3992	.5411	.556	53.9	47.08	17.32
6	Georgia	B. S.	435	14963	.1222	.605	181.8	76.21	23.75
7	Connecticut	B. S.	450	16375	.1157	.682	179.7	76.83	25.00
8	North Carolina	Arm. Cruiser	502	14570	.1562	.585	115.2	72.88	25.07
9	St. Louis	Prot. Cruiser	424	9665	.1957	.610	126.8	66.00	22.50
10	Rodney	B. S.	325	9690	.1309	.660	282.3	68.00	26.70
11	Narkeeta	Tug	92.5	190	1.951	.583	240.0	20.05	7.92
12	Sheadle	Lake Fgt.	530	13303	.1884	.859	89.3	56.00	18.46
13	Tremont	Ocean Fgt.	490	11410	.2004	.782	97.0	58.00	19.21

It will be observed that in any particular case N is proportional to $\dfrac{I}{V^3}$, and hence is proportional to the reciprocals of the Admiralty coefficients, which are both proportional to $\dfrac{V^3}{I}$.

The Extended Law of Comparison method of estimating power, though better than the Admiralty Coefficient method, is essentially but an improved form of the latter.

The assumption that all resistances follow the Law of Comparison is in error as regards the Skin Resistance. This tends to make us overestimate when powering a large ship from the results of a small ship, and *vice versa.* The efficiency of propulsion is not constant, and the efficiency of the new ship may be different from that of the old ship. This source of error is common to all methods of estimating power from trial results.

We have seen that resistance is materially affected by variation of the displacement length coefficient and of the longitudinal coefficient. The method of the Extended Law of Comparison takes no account directly of such variations and is subject to error accordingly. In fact, curves of N, as in Fig. 276, are of very little value without full information as to the ships to which they refer. Thus, suppose we wish to power a ship for which $\frac{V}{\sqrt{L}}$ is to be .8. For this speed length ratio we find in Fig. 276 values of N which vary radically. Thus Nos. 6 and 7 would give $N = .275$. There are a number of other values between .30 and .35. Nos. 4 and 5 would give .475, while No. 3 would give .72. These values are thoroughly discordant. It is evidently desirable, when powering a new ship, to use curves of N from ships of the same type having approximately the same longitudinal and displacement length coefficients.

3. **Powering Sheet using Extended Law.** — If a number of speed and power curves of various types of ships are available, their practical use in powering is materially facilitated by reducing them to curves of N, as in Fig. 276, but plotting these curves as in Fig. 277. A large sheet should be used, section ruled vertically with lines representing equal intervals of longitudinal coefficient and horizontally with lines representing equal intervals of $D \div \left(\frac{L}{100}\right)^3$ as indicated.

Curves of N are placed upon this sheet so that the termination corresponding to the maximum trial speed is located at the point corresponding to the longitudinal coefficient and the $D \div \left(\frac{L}{100}\right)^3$ for the ship. All curves terminate at their other extremity where $\frac{V}{\sqrt{L}} =$ .5, and a vertical line is drawn down from this extremity to the point where $N = 0$ or has a given value. For greater clearness each curve is numbered, and the corresponding spot where $N = 0$, $\frac{V}{\sqrt{L}} = .5$ is marked O with a subscript number the same as the curve number. When for the datum point $\frac{V}{\sqrt{L}} = .5$ but N is not 0, the value of N is indicated. The same scales of $\frac{V}{\sqrt{L}}$ and of N are used for all curves, and being drawn upon a separate piece of tracing cloth can be adjusted over or under the main sheet so as to apply to any curve. Thus, in Fig. 277 the dotted lines represent the scale in position for No. 6 curve of N. When reliable data is available for but a single spot — not a curve — it may be located on the powering sheet as the spots marked 10, 11, 12 in Fig. 277. Each spot must have its o spot also located as shown.

When powering a new design we will know the values we expect to use for longitudinal coefficient, for $D \div \left(\frac{L}{100}\right)^3$ and for $\frac{V}{\sqrt{L}}$. Locating on the sheet the point corresponding to the longitudinal coefficient and the value of $D \div \left(\frac{L}{100}\right)^3$, it is obvious that the best curves of N to use are those terminating nearest to the located spot. Having selected the curves of N to be used, adjust the scale to the chosen ones in succession and from each curve take the value of N corresponding to the $\frac{V}{\sqrt{L}}$ for the new design. Sometimes there may be reasons for giving more weight to some of those values of N than to others. If not, the average of the values of N is the proper value to use for the new design, as a basis for an estimate of the neat power and the variation in the various values will assist in fixing the margin of power which should always be allowed over and above the neat estimated power. In practice several values of N should be taken from each available curve corresponding to definite values of $\frac{V}{\sqrt{L}}$ and an estimated curve of I.H.P. determined extending above and below the intended speed of the new design.

The few curves of N in Fig. 277 are shown simply to indicate how a working sheet should be prepared. Such a sheet should have a large number of curves on it, the more the better, but no curves or spots should be used which are not derived from reliable results of careful trials. Published trials are not always reliable.

The advantage of a powering sheet laid off as shown in Fig. 277 is that when a designer is considering a question of powering it enables him to determine immediately whether his power data from previous ships is applicable to the case or whether he is working in a region not covered by reliable data in his possession.

The error arising from the application of the results of a small ship to the powering of a large ship can be approximately corrected if estimates of the frictional effective horse-power at corresponding speeds of the two are made. By applying the Law of Comparison to the frictional effective horse-power of the small ship and deducting from the result the frictional effective horse-power of the large ship we determine the error in the effective horse-power incident to the use of the Extended Law of Comparison, and the error in the indicated horse-power will usually be about double that in the effective horse-power. By an obvious similar method we can correct when passing from a large to a small ship.

The error due to variation of propulsive efficiency from ship to ship is not great when we use results of similar ships with somewhat similar types of propelling machinery. But caution should be exercised and liberal margins allowed if, for instance, we wish to power a turbine vessel and have available only data from vessels with reciprocating engines.

The main difficulty with the Extended Law of Comparison method as a practical working proposition is the fact that few or no designers will have available reliable trial results which will cover the whole field of speed, longitudinal fineness and displacement length coefficient.

39. Standard Series Method

1. Use of Standard Series Results. — In addition to the comparatively simple methods of powering ships described there have been many others proposed which are as a rule more complicated. Many involved formulæ for resistance have been brought forward from time to time.

Skin resistance is readily estimated by a formula using the coefficients of Froude and Tideman, but no general formula giving residuary resistance accurately for any wide range of speed, proportions, and fullness of model has yet been brought forward. We have seen that the best approximate methods of powering hitherto used are all weak in leaving largely to the skill and judgment of the designer, to his guesswork, the effect of proportions and fullness of model, and that in order to make satisfactory guesses the designer must have an accumulation of data possessed by few.

Now by the use of the data given in Figs. 78 to 120 it is possible to estimate with great accuracy the effective horse-power of a ship of any displacement, dimensions, and longitudinal coefficient upon the lines of the Standard Series. Furthermore, such a curve of effective horse-power will approximate fairly closely the E.H.P. of models upon different lines. For with displacement, length, midship section area, and longitudinal coefficient fixed, any variations in shape that would be made in good practice will have a comparatively minor effect upon resistance. Hence, with the aid of the Standard Series the problem of powering a ship is solved in two steps.

First: From the Standard Series results get out a curve of E.H.P. for a ship of the same displacement, length, beam draught ratio and longitudinal coefficient.

Second: From the E.H.P. estimate the I.H.P. by applying a suitable coefficient of propulsion.

When following this method there are two principal sources of error.

First, there is the possibility that the lines used may differ so much from those of the Standard Series that the estimated E.H.P. may be materially in error. This source of error may be avoided by closely following the lines of the Standard Series unless lines positively known to be superior are available.

Second, the coefficient of propulsion chosen may be in error. This is an unavoidable source of error, and it is on this point only that the designer, when using the Standard Series method, must use some guesswork.

2. Propulsive Coefficients to Use. — When an accumulation of power data is not available, it is generally safe, when using lines closely resembling those of the Standard Series, to assume a nominal efficiency of propulsion in the vicinity of 50 per cent based upon indicated horse-power for reciprocating engines and somewhat less, say 46 per cent, for the usual run of turbine jobs, but using shaft horse-power in this case. These average efficiencies are based upon the E.H.P. of the bare hull and are sufficiently low to allow for the average run of appendages.

The above is independent of accumulated data of experience and will enable fairly good results to be obtained without such data, but when such is available it should be made use of to the fullest extent.

Thus, if we have a reliable speed and power curve of a vessel, we can estimate from the Standard Series the E.H.P. for a vessel on Standard Series lines having the same displacement, length, area of midship section and ratio of beam to draught. Then from the I.H.P. curve of the actual vessel we can determine the nominal efficiency of propulsion. The same nominal efficiency should be found for another vessel of the same general type as the vessel whose trial results are known, including type of engines and propellers. Or it may be that there is some change made in the new vessel which leads us to anticipate a certain reduction in nominal efficiency of propulsion. Knowing the old nominal efficiency and the probable reduction, the new nominal efficiency to be expected is determined.

Any one who finds from reliable data of a given type of vessel a nominal efficiency of propulsion below 50 per cent, should be careful when powering a new vessel of the type to use the nominal efficiency based upon preceding results. Analysis of trial results by the aid of the Standard Series will disclose plenty of nominal efficiencies below 50 per cent. They may be due to lines inferior to those of the Standard Series, to inefficient propelling machinery, or to inaccurate power data. All trials are not handled so that the resulting speed and power data will be accurate. Still nominal efficiencies of propulsion of 50 per cent for indicated horse-power of reciprocating engines and 46 per cent for shaft horse-power of turbines are often materially exceeded, and when it is found that they have not been reached endeavor should be made to locate the trouble.

3. Advantages of Standard Series Method. — The Standard Series method of estimating power has the great advantage that even if the resistance of a given ship is different from the corresponding Standard Series ship the variations of resistance with varying dimensions and shape of ships of the type will follow closely the variations deduced from the Standard Series. In other words, the Standard Series may be used as a reference scale to determine relative resistances of ships of constant type of any dimensions and proportions. A tape measure need not be accurate to determine the ratio of two lengths, and even if from the Standard Series curves we cannot accurately estimate *a priori* the resistance of a ship of a given type, we can estimate with fair accuracy the ratio of the resistances of two ships of the type; and if we have accurate power data for one or more such ships we can use it to establish the proper nominal efficiency of propulsion from which, using the Standard Series, we can estimate with ample accuracy the power required for other vessels of the type. For this purpose it makes no difference whether the nominal efficiency is or is not the real efficiency of propulsion. If it is really typical of the type of vessel in hand it is adequate for powering purposes.

The fact that the nominal efficiency of propulsion, which does not vary much without good reason, is the only quantity which must be estimated or guessed at from experience is much in favor of the Standard Series method. Furthermore, in the great majority of cases the efficiency of propulsion does not vary much in the vicinity of full speed.

Hence, for practical purposes for use in future designs, we can characterize a complete trial by a single number, namely, the efficiency of propulsion whether actual or nominal. This is a great advantage where there is a mass of data to deal with. In the Standard Series method of powering all other factors are taken care of by the method, automatically, as it were.

While by the Standard Series method estimates of power are much simplified and should be made with more accuracy than by any of the other methods of approximation described, they are still estimated, and the designer should be careful always to allow a margin of power adequate to the necessities of the case. By any conceivable method of powering two sister ships would be given the same power for the same speed, yet sister ships do not always develop the same power on trial and do not always make the same speed for the same power. Changes from previous vessels made with a view to improvement sometimes turn out badly. Propellers frequently disappoint the designer, and the quick running propellers required by turbines are especially uncertain.

The designer who is an optimist in choosing the efficiency of propulsion to be expected may be very pessimistic after the trial. The time for pessimism is when the powering is being done, not when the trial is being run.

INDEX

	PAGE
Admiralty coefficient method, powering ships	294
Air disengaged around moving ships	64
Air friction, Zahm's experiments	82
Air resistance	82
Air resistance of planes	84
Air resistance of ships	86
Appendages, allowance for, in powering ships	126
Appendages fitted on ships	123
Area, coefficients of, for elliptical blades	134
Area, developed, determination of	132
Area of midship section, effect upon resistance	97
Area of propeller blades, effect of	173
Atlantic liner, design of propeller for	248
Augmented surface method, Rankine's	291
Babcock, measurements of settlement of ships in channels	120
Back of blade	128
Beam and draught, effect of ratio upon resistance	96
Beaufoy's eddy resistance experiments, John's analysis of	69
Bilge keels, resistance of	123
Blade area, effect of	173
Blade, back, variation of pitch over	159
Blade sections, propeller, strength of	223
Blade thickness, correction factors for	179
Blade thickness, effect of	172
Blade thickness ratio or fraction	130
Blades, detachable, connections of	239
Blades, propeller, forces on	216
Blades, propeller, moments on	216
Blades, propeller, number of	245
Blades, propeller, stresses allowed	235
Bow, change of level under way	108
Bow, shape of, effect upon resistance	93
Cargo vessel, design of propeller for	252
Cavitation, accepted theory inadequate	183
Cavitation, cause of	188
Cavitation causes failure of Law of Comparison	151
Cavitation, cure for	192
Cavitation, effect of broad blades upon	190

	PAGE
Cavitation, experiments with narrow and broad blades	191
Cavitation, model experiments with	186
Cavitation, nature of	182
Cavitation, possible methods of experimental investigation	185
Cavitation, possible theories of	184
Cavitation, theory of	188
Cavitation, visible phenomena in model experiments	188
Centrifugal force, stresses due to	226
Channels, shallow, settlement of ships under way	120
Coal endurance estimated from surface	292
Coefficients of propeller performance, characteristic	161
Comparison, Law of, deduction	26
Compressive stresses on propeller blades	224
Current, tidal, elimination of effect on trials	267
Depth for no change of resistance	116
Depth of various trial courses	118
Design of propeller for Atlantic liner	248
Design of propeller for destroyer	250
Design of propeller for gunboat	251
Design of propeller for large cargo vessel	252
Design, $\rho\delta$ diagrams of	176
Design of propellers, reduction of model experiment results for	164
Destroyer, design of propeller for	250
Detachable blades, connections of	239
Developed area, determination of	132
Deviations of shafts, actual and virtual	211
Deviations of shafts, virtual, due to motion of water	213
Diagrams, $\rho\delta$, for design	176
Diameter ratio defined	130
Dimensional formulæ	33
Direction of rotation of propellers	245
Disc area	132
Disc area ratio	132
Displacement length ratio, influence on resistance	104
Displacement length ratio defined	99
Disturbance of water by a ship	50
Draught and beam, effect of ratio upon resistance	96
Docking keels, resistance of	123
Eddy resistance, formulæ for, inclined plates	71
Eddy resistance, formulæ for, normal plates	70
Eddy resistance, formulæ for practical use	72
Eddy resistance, limitations of rear suction formula	72
Effect of foulness upon skin resistance	66
Efficiency, effect of shaft inclination upon	214
Efficiency, hull	197
Efficiency, maximum attainable in practice	177

	PAGE
Efficiency, maximum of $\rho\delta$ diagrams	177
Efficiency of a propeller, general considerations	138
Efficiency of ideal propellers	153
Efficiency of propulsion, determination from trial results	289
Efficiency of propulsion, values for practical use	301
Efficiency, propeller, deduction from experimental results	160
Elliptical blades, coefficients of area for	134
Endurance, coal, estimated from surface	292
Expanded area	133
Extended Law of Comparison, powering ships	295
Face of blade	128
Feathering paddle wheels	256
Float area of paddle wheels	259
Flow past vessel, lines of	52
Focal diagrams	48
Forces on propeller blades	216
Four-bladed propellers, design from $\rho\delta$ diagrams	180
Four-bladed propellers, ratios connecting with three-bladed	180
Friction, initial	279
Friction, initial, determination of	281
Friction in propeller action and head resistance	144
Friction, load	279
Froude, R. E., skin resistance constants	62
Froude's Law	26
Froude's propeller theory, formulæ from	140
Froude, W., skin resistance experiments	58
Gaillard's experimental investigations of trochoidal waves	19
Girth parameters	40
Girths of sections	40
Greenhill's propeller theory, formulæ from	143
Groups of waves	17
Gunboat, design of propeller for	251
Havelock's wave formulæ	55
Hovgaard's observations of wave patterns	55
Hub, propeller, effect of size	169
Hubs, propeller, fair waters to	125
Hull efficiency	197
Humps and hollows of resistance	78
Immersion of propellers, effect on efficiency	210
Inclination of propeller blades, effect of	168
Inclinations of shafts, effect upon efficiency	214
Inclination of shafts, effect upon vibration	214
Inclination of shafts, virtual, due to motion of water	213
Indicated horse-power, components of	279

	PAGE
Indicated thrust	284
Initial friction	279
Initial friction, determination of	281
Jet propulsion	26c
Joessel's eddy resistance results	68
John's analysis of Beaufoy's eddy resistance results	69
Keels, bilge, resistance of	123
Keels, docking, resistance of	123
Kelvin's wave patterns	53
Kirk's method for powering ships	292
Law of Comparison, application to centrifugal fans	31
Law of Comparison, application to propellers	31
Law of Comparison, application to ship's resistance	30
Law of Comparison, application to steam engines	30
Law of Comparison applied to propellers	150
Law of Comparison, deduction	26
Law of Comparison, formulæ for simple resistances which follow	32
Law of Comparison, not applicable to skin resistance	63
Leading edges of propellers, fluid pressures at	194
Length, effect upon resistance	98
Length, illustration of influence upon resistance	105
Level of vessel, change under way	52
Level of water around vessel, change under way	52
Lines of flow over vessel	52
Load friction	279
Location of propellers	241
Longitudinal coefficient, effect upon resistance	97
Longitudinal coefficient, influence on resistance in Standard Series	103
Luke's experiments on wake fractions and thrust deductions	198
Margin to be allowed in powering ships	303
Material of propellers	247
McEntee, limits of propeller efficiency	153
Measured courses, desirable features of	262
Measured miles, desirable features of	262
Midship section area, effect upon resistance	97
Midship section area, optimum for resistance	104
Midship section coefficient, effect on resistance	95
Midship section shape, effect on resistance	95
Model basin methods	87
Model propeller, analysis of experimental results	158
Model propeller experiments, reduction for design work	164
Model propeller, plotting experimental results	157
Model propellers, experimental methods	155
Model trial results applied to determine ship's resistance	88

INDEX

	PAGE
Moments on propeller blades	216
Motion past a ship, differences from ideal	51
Nominal pitch	158
Nominal slip	158
Number of blades of propellers	245
Number of propeller blades, effect of	167
Number of propellers	241
Obliquity factor for wetted surface	38
Obliquity of flow of water at propeller	215
Obliquity scales for wetted surface calculations	38
Paddle propulsion	254
Paddle wheel location	257
Paddle wheels, dimensions and proportions	258
Paddle wheels, feathering	256
Paddle wheels, float area	259
Paddle wheels, number of blades	260
Paddle wheels with fixed blades	255
Parallel middle body, curves for finding resistance of vessels with	107
Parallel middle body, experiments on models with	106
Parallel middle body, optimum percentages for resistance	107
Parameters, girth	40
Parent lines, derivation of models from	91
Pitch angle	129
Pitch, decreasing	129
Pitch, increasing	129
Pitch, nominal	158
Pitch of back of blade	129
Pitch of helicoidal surface	128
Pitch ratio defined	129
Pitch ratio, effect upon propeller action	171
Pitch, variation over back of blade	159
Pitch, variation of, for twisted blades	135
Pitch, virtual	158
Pitch, virtual, determination from experimental results	159
Planes, resistance in air	84
Plane, thin, flow past	67
Powering ships, Admiralty coefficient method	294
Powering ships, allowance for appendages	126
Powering ships, extended Law of Comparison	295
Powering ships, Kirk's method	292
Powering ships, Rankine's method	291
Powering ships, Standard Series method	300
Practical application of model propeller results	175
Pressure, fluid, at leading edges of propellers	194

	PAGE
Progressive speed trials	264
Progressive trials, accuracy of results attainable	265
Progressive trials, conditions of	278
Progressive trials, methods of conducting	271
Projected area	132
Propeller action, comparison between theories and experience	147
Propeller action, comparison of theories	143
Propeller action, formulæ on various theories	146
Propeller action, friction and head resistance	144
Propeller action, Froude's theory of	138
Propeller action, Greenhill's theory of	138
Propeller action, Rankine's theory of	138
Propeller action, theories of	136
Propeller, area of	132
Propeller blade sections, strength of	223
Propeller blades, forces on	216
Propeller blades, moments on	216
Propeller blades, stresses allowed	235
Propeller blades, width of	247
Propeller bossing or spectacle frames, resistance of	126
Propeller, coefficients of performance, characteristic	161
Propeller, delineation of	131
Propeller design, reduction of model experiment results for	164
Propeller efficiency, deduction from experimental results	160
Propeller efficiency, ideal	153
Propeller efficiency, McEntee's limits of	153
Propeller immersion, effect on efficiency	210
Propeller, obliquity of water flow at	215
Propellers, location of	241
Propellers, material of	247
Propellers, model, analysis of experimental results	158
Propellers, model experimental methods	155
Propeller, model, plotting experimental results	157
Propellers, number of	241
Propeller suction	207
Propulsion by jets	260
Propulsive coefficients, values for practical use	301
Propulsive efficiency, determination from trial results	289
$\rho\delta$ diagrams for practical use	176
Rake of propeller blades, effect of	167
Rake ratio	130
Ram bow, effect upon resistance	93
Rankine's augmented surface method	291
Rankine's propeller theory, formulæ from	140
Rayleigh's formula for eddy resistance	67
Residuary resistance, analysis of curves	80
Residuary resistance, curves of	79

INDEX

	PAGE
Residuary resistance from Standard Series	101
Residuary resistance, method of plotting for analysis	90
Resistance, air	82
Resistance, air, defined	58
Resistance, air, of ships	86
Resistance coefficients and constants, variables used in plotting	98
Resistance, depth for no change	116
Resistance, disengaged air, effect upon	64
Resistance, eddy, defined	57
Resistance, eddy, formulæ for	67
Resistance, eddy, of inclined plane	67
Resistance, factors affecting	92
Resistance, increased in rough water	121
Resistance in shallow water, percentage variations	118
Resistance, kinds of	57
Resistance of ship, deduction from model results	88
Resistance, residuary, analysis of curves	80
Resistance, residuary, curves of	79
Resistance, residuary, from Standard Series	101
Resistance, residuary, method of plotting for analysis	90
Resistance, shallow-water effects	112
Resistance, skin and wave, relative importance	58
Resistance, skin, defined	57
Resistance, skin, determination for ships	99
Resistance, skin, effect of foulness	66
Resistance, skin, Law of Comparison not applicable	63
Resistance, skin, of ships, deduced from plane results	61
Resistance, skin, R. E. Froude's constants	62
Resistance, skin, Tideman's constants	63
Resistance, skin, variation of coefficients	60
Resistance, skin, W. Froude's experiments	58
Resistance, wave, defined	57
Rota's experiments on depth and resistance	116
Rotation of propellers, direction of	245
Rough water, reduction of speed in	121
Screw, true	128
Sections, girths of	40
Settlement in shallow water	119
Settlement of ships in shallow channels	120
Shaft bossing, effect on wake	205
Shaft brackets, effect on wake	205
Shaft deviations, actual and virtual	211
Shaft deviations, virtual, due to motion of water	213
Shaft inclination, virtual, due to motion of water	213
Shallow water, changes of trim and settlement	119
Shallow water, effect upon resistance	112
Shallow-water resistance, percentage variations	118

	PAGE
Shape of bow and stern, effect upon resistance	93
Shape of midship section, effect upon resistance	95
Sink and source motion	4
Sink and source motion in uniform stream	5
Skin resistance determination for ships	99
Slip angle	130
Slip-angle values	148
Slip, effect of, upon propeller action	174
Slip, nominal	158
Slip of paddle wheels	255
Slip percentage	131
Slip ratio	131
Slip, variation because of shaft inclination	211
Slip, virtual	158
Spectacle frames or propeller bossing, resistance of	126
Speed and power trials, general considerations	264
Speed of advance	130
Speed of propeller	130
Speed of slip	130
Speed ratio	131
Speed, reduction in rough water	121
Speed trials, progressive	264
Squat in shallow water	119
Squat under way	108
Standard Series method of powering ships	300
Standard Series method of powering, advantages of	302
Standard Series method of powering, margin to allow	303
Standard Series of model propel'ers	170
Standard Series, residuary resistance from	101
Stanton's eddy resistance results	70
Steady motion formula	1
Steady motion formula, failure of	3
Steady motion past ships	2
Stern, change of level under way	108
Stern, shape of, effect upon resistance	93
Stream forms	6
Stream lines	1
Stream lines around sphere	10
Stream lines past elliptical cylinders	7
Stream motion past a ship, ideal	51
Strength of propeller blade sections	223
Stresses allowed in practice on propeller blades	235
Stresses, compressive, on propeller blades	224
Stresses due to centrifugal force of propellers	226
Stresses, tensile, on propeller blades	226
Struts, resistance of	124
Suction of propellers	207
Superposition of trochoidal waves	17

INDEX 313

	PAGE
Tensile stresses on propeller blades	226
Thickness of propeller blades, effect of	172
Thrust deduction	197
Thrust deduction, approximate determination of	200
Thrust deduction coefficient	197
Thrust deduction, determination from trial results	290
Thrust deduction factors	197
Thrust deduction, variation of	198
Thrust, indicated	284
Tidal current, elimination of effect on trials	267
Tideman's skin resistance constants	63
Trials, progressive, conditions of	278
Trials, progressive, methods of conducting	271
Trial results, analysis of	279
Trim, change of, in shallow water	119
Trim, change of, under way	108
Trim, effect upon resistance	94
Trim of vessel, change of, under way	52
Trochoidal theory of waves, applicability of	18
Trochoidal theory of waves, Gaillard's investigations	19
Trochoidal wave theory	11
True screw	128
Twisted blades	135
Two-bladed propellers, ratios connecting with three-bladed	181
Vibration, effect of shaft inclination upon	214
Virtual pitch	158
Wake, components of	195
Wake, effect of shaft brackets on	205
Wake factor	197
Wake fraction, approximate determination of	200
Wake fraction, determination from trial results	290
Wake fraction, estimates from trial results	201
Wake fractions	196
Wake fractions, variation of	198
Wake, frictional	195
Wake, how it affects propulsion	196
Wake, percentage, Froude's expression	197
Wake, stream line	195
Wake, wave	195
Wave formulæ, Havelock's	55
Wave formulæ, Kelvin's	53
Wave groups	17
Wave patterns, Hovgaard's observations	55
Wave patterns, Kelvin's	53
Wave patterns of ships	55

	PAGE
Wave resistance	73
Wave resistance, general formula for	76
Wave resistance, humps and hollows in	78
Waves dimensions of actual	23
Waves, energy of trochoidal	15
Waves, formulæ for trochoidal	13
Waves of translation	23
Waves, relation to wind causing them	24
Waves, shallow water, trochoidal	21
Waves, solitary	21
Waves, superposition of trochoidal	17
Waves, trochoidal	11
Wave system, resultant	74
Wave systems, bow and stern	73
Wetted surface calculations, correction factors for	40
Wetted surface calculations, form for	39
Wetted surface coefficients, average	46
Wetted surface coefficients, variation of	44
Wetted surface, factors affecting	47
Wetted surface, formulæ for	43
Wetted surface, obliquity factor for	38
Wetted surface of appendages	37
Width of propeller blades	247
Winds, relation to waves produced	24
Yorktown model results compared with Standard Series	103
Zahm's experiments on air friction	82

www.ingramcontent.com/pod-product-compliance
Lightning Source LLC
Chambersburg PA
CBHW050857300426
44111CB00010B/1286